AF846997

Fluid Mechanics and Its Applications

Volume 115

Series editor

André Thess, German Aerospace Center, Institute of Engineering Thermodynamics, Stuttgart, Germany

Founding Editor

René Moreau, Ecole Nationale Supérieure d'Hydraulique de Grenoble, Saint Martin d'Hères Cedex, France

Aims and Scope of the Series

The purpose of this series is to focus on subjects in which fluid mechanics plays a fundamental role.

As well as the more traditional applications of aeronautics, hydraulics, heat and mass transfer etc., books will be published dealing with topics which are currently in a state of rapid development, such as turbulence, suspensions and multiphase fluids, super and hypersonic flows and numerical modeling techniques.

It is a widely held view that it is the interdisciplinary subjects that will receive intense scientific attention, bringing them to the forefront of technological advancement. Fluids have the ability to transport matter and its properties as well as to transmit force, therefore fluid mechanics is a subject that is particularly open to cross fertilization with other sciences and disciplines of engineering. The subject of fluid mechanics will be highly relevant in domains such as chemical, metallurgical, biological and ecological engineering. This series is particularly open to such new multidisciplinary domains.

The median level of presentation is the first year graduate student. Some texts are monographs defining the current state of a field; others are accessible to final year undergraduates; but essentially the emphasis is on readability and clarity.

More information about this series at http://www.springer.com/series/5980

Zeka Mazhar

Fully Implicit, Coupled Procedures in Computational Fluid Dynamics

An Engineer's Resource Book

Springer

Zeka Mazhar
Department of Mathematics
Eastern Mediterranean University
Turkish Republic of Northern Cyprus
Turkey

ISSN 0926-5112 ISSN 2215-0056 (electronic)
Fluid Mechanics and Its Applications
ISBN 978-3-319-29894-8 ISBN 978-3-319-29895-5 (eBook)
DOI 10.1007/978-3-319-29895-5

Library of Congress Control Number: 2016931322

© Springer International Publishing Switzerland 2016
This work is subject to copyright. All rights are reserved by the Publisher, whether the whole or part of the material is concerned, specifically the rights of translation, reprinting, reuse of illustrations, recitation, broadcasting, reproduction on microfilms or in any other physical way, and transmission or information storage and retrieval, electronic adaptation, computer software, or by similar or dissimilar methodology now known or hereafter developed.
The use of general descriptive names, registered names, trademarks, service marks, etc. in this publication does not imply, even in the absence of a specific statement, that such names are exempt from the relevant protective laws and regulations and therefore free for general use.
The publisher, the authors and the editors are safe to assume that the advice and information in this book are believed to be true and accurate at the date of publication. Neither the publisher nor the authors or the editors give a warranty, express or implied, with respect to the material contained herein or for any errors or omissions that may have been made.

Printed on acid-free paper

This Springer imprint is published by SpringerNature
The registered company is Springer International Publishing AG Switzerland

Preface

I was a graduate student in the Department of Computer Science at the University of Waterloo in 1977. One day, Prof. Bill Nicoll, whom I later learned was one of the founders of the University and of the Department of Mechanical Engineering, knocked on my door with his ever smiling face and humble posture. I could not know this would be a turning point in my life.

He told me he had seen my technical note published by the University on the finite element solution of the convective diffusion equation. He then asked if I would like to join his team in the Mechanical Engineering Department and continue my Ph.D. studies there. He also mentioned his plans to visit Prof. D.B. Spalding in London to do a joint study in computational fluid dynamics, and asked if I would be able to accompany him. His offer came as a great surprise, since, as I explained, I had no knowledge of fluid dynamics. I was only a mathematician with some specialization in numerical analysis. He assured me that this would not cause a problem since I would be working on the numerical aspects of the subject. He added that I would also improve myself in due course by taking certain graduate courses in fluids. He gave me one of his latest papers and asked me to examine the partial differential equations in the paper and contact him later.

After I examined the paper I decided to accept the offer, but there was a serious problem. At the time I was newly wed and my wife, Ümit, was expecting our first child in a couple of months; therefore, I had to discuss the issue with her. To my surprise she accepted the idea without hesitation. Soon, I visited Prof. Nicoll in his office and told him of my decision. I expected that it would take some time for my transfer to another faculty, but right away he took me to my new office, which it was already prepared, and insisted that I start work immediately.

Sometimes after my transfer, I learned that the principal of the University initially had rejected my transfer based on his belief that a mathematician would not be successful in engineering. But Prof. Nicoll had insisted that if engineers were successful in mathematics, mathematicians could well succeed in engineering, thus initiating my career in the field of computational fluid dynamics.

Unfortunately, our collaboration with Prof. Nicoll ended after his sudden and unexpected death a couple of months later.

The second turning point in my life was when I met another wonderful man, Prof. George D. Raithby, with whom I worked closely for about four fruitful years until completion of my Ph.D. thesis. He was more than a supervisor to me. He was a friend who insisted being called 'George'; something that, out of respect, I initially had difficulty in doing. Together we completed a comprehensive analysis of the available computational procedures for the solution of incompressible fluid flow problems. We came up with some new ideas on improving the performance of these procedures.

In appreciation of his valuable help, I studied hard to complete my thesis. I well remember that on one occasion, while working late in my office, that the University security officer came and asked me to "ring my poor wife at home as she was worried about my whereabouts."

All the hard work paid off as we came up with a good job which was accepted without any amendment by the examiners. After I finished my Ph.D., Prof. Raithby told me that this was the first time, in the history of the University that a Ph.D. thesis was accepted with no amendments whatsoever. The key to my success was, without doubt, his gentle and kind encouragement and guidance.

I would have dearly liked to have continued working with him to realize the ideas I then had in my mind on the development of new schemes on the subject; of which, arising from my unique situation as a numerical analyst, the feeling that a procedure which would directly make use of the most naive form of the discretized governing equations may work better. After all, as Albert Einstein's said "God does not play dice." In fact, every physical process in nature works in a superbly designed mathematical mechanism. Therefore, why do we try to make physically unrealistic assumptions about nature, instead of making use of this nice formulation in its primitive form?

Unfortunately, my new adventure in founding the Eastern Mediterranean University (EMU) in my native country, Cyprus, prevented me from continuing to work on this subject, at least for some time. This was mainly due to the lack of computing machines at EMU such as those at the University of Waterloo.

It was only after the introduction of personal computers at EMU in the early 90s that I was able to return to the subject and start to implement some of these ideas by directly attacking the primitive form of the equations. In the meantime, a few papers were published by Prof. Raithby and his colleagues. They were reporting that keeping more effects of the coupling between the pressure and velocity fields significantly decreased the need for high under-relaxation by enhancing the range of convergent relaxation parameters. I shared their belief. Surprisingly, however, the overall convergence of the proposed procedures was not much of an improvement on the segregated ones.

Unfortunately, after these findings were published, researchers reverted to the old procedures and thus research in this regard was terminated. There was an unspoken acceptance that any attempts to solve the equations in their primitive form were to no prevail.

I had the feeling that something was going wrong, not with the basic idea, but with the solution procedures themselves. Otherwise, considering the indications, the performances would have been much better. I encouraged one of my Master students, Türkan Güngörmüş, to work with me to understand this dilemma, but our efforts were not fruitful and resulted in similar conclusions.

I did not give up. After further work on the subject, I realized that the main culprit was sneaking in the reordering of the equations. In fact, the reordering applied was worsening the diagonal dominancy of the coefficient matrix, thus slowing the convergence of the iterative process.

Based on this observation, I decided to accept the primitive form as it is, and try some new ideas on the solution algorithm. A natural equation for pressure lay there in the equations, but the main diagonal sub-matrix for pressure was null. Although it was possible to collect nonzeros to the sub-matrix, it would result in a Poisson form, which I would not prefer. Then, I noticed that an incomplete decomposition of the primitive matrix helps, not perhaps in creating nonzero terms in the null part of the original coefficient matrix, but in the main diagonal coefficients of the sub-matrices of the incomplete decomposition. That would be enough for me.

It was not really easy to see what the underlying equations were, so I had to work on long sheets of joined-up squared papers. I multiplied the decomposition matrices and equated them to the primitive matrix. There it was. I had a few nice-looking equations that would give me what I was looking for.

It was a quick implementation of the algorithm that gave me the actual surprise. The algorithm had worked well; it was converging twice as fast when compared to the best segregated procedure. Furthermore, the range of convergent relaxation parameters was broadly widened. This was like replacing my old car by one with a turboprop.

The most satisfying side of this achievement was that at last we got rid of the Poisson-type equations.

It was only then that I went back and examined what the mechanism of the derived equations was doing to invoke such a remarkably well-working procedure.

I realized that the mechanism engineered by the equations not only brings together the effects of the coefficients from the velocity field, but also all of the effects of the mass conservation constraint and all of the effects of the coefficients of the pressure terms in the momentum equations.

After the publication of my first paper on the subject, I concentrated on the improvement of the algorithms by trying various kinds of placements of the vectors in the incomplete decompositions. This turned out to be fruitful, from which came the procedures named FICS (Fully Implicit Coupled Procedure) in this book. The procedures converged about 20 times faster than the segregated procedures and convergence was achieved even without any relaxation. Although the derivation of the relevant formulas was not straightforward, the derived formulas were so simple that they allowed an easy implementation of the solution process.

Following this achievement, I decided to share my findings with other researchers. It was not possible to fully express myself in a journal paper form, so I decided to write a book about the subject.

While working on the manuscript of the book, I was still skeptical on the actual physical reason for the overwhelming performance of the procedures. Numerically, there was no problem and everything was working much finer than I would ever dream of. However, there should have been a meaningful physical explanation to this as well. Finally, I realized that the procedures were actually centralizing on the most fundamental flow property of incompressible fluids.

This important property, which was unfortunately overlooked in the past decades by researchers, including myself as well, was in fact, one of the fundamental rules of fluid flow.

A fluid particle may only move from one point to another if there exists a pressure difference between the two points. The movement is initiated and dictated by the pressure differences characterized by the pressure derivative terms in the momentum equations. While this movement is underway, the particle's velocity components must obey the mass conservation constraint. Thus, the effects of the pressure derivative terms in the momentum equations and those of the velocity derivative terms in the mass conservation equation are equivalently important. These effects are related directly to the fluid flow, but not to the physical fluid properties governed by the diffusion and convection terms.

Therefore, in order that a solution procedure be more efficient, instead of concentrating on the convection and diffusion terms, as was done with the segregated procedures, it should have concentrated on the correct imposition of the pressure changes as well as the mass conservation constraint in all stages of the solution process. The physical properties could perhaps be taken care of in due course.

To put it more precisely, the effect of the pressure terms, in parallel to the mass conservation principle, should have been regarded as a constraint, perhaps to be better named as "pressure conservation constraint."

The mechanism engineered by the new family of procedures was in fact, incorporating this new viewpoint into the solution process. Consequently, this constituted an efficient remedy for all of the problems that were encountered in the classical segregated procedures. The endeavor of creating an explicit pressure Poisson equation in these procedures was actually disrupting the vitally important coupling between the velocity and pressure fields.

Presentations in journal article form do not contain any details, but only vast amounts of references to previously published articles. Consequently, a novice engineering graduate or a field engineer needs to spend months researching these multifarious papers in order to grasp the details involved in the solution processes. These efforts usually end up in a complete frustration and failure to actually implement the procedures.

Books written in this area are usually engineering based and do not address mathematically oriented researchers. They do not either contain adequate information regarding the computer implementation of the procedures. As a mathematician, I accept that some knowledge in the engineering side helps, but when it comes to the development of better numerical algorithms, engineering matters become less important.

Postgraduate books written by mathematicians, on the other hand, are often too general. They are too much theoretically oriented and consequently suffer from heavy mathematical notation. As a result, engineers without deep mathematical theory cannot make direct use of the material.

I always recall and believed in Prof. Nicoll's thoughts; mathematicians with a good knowledge of numerical analysis have much to contribute to the field but always in collaboration with engineers. From such a collaboration more fruitful results will emerge.

On the other hand, it is unfortunate that an experienced researcher or an application specialist is usually reluctant to try and adapt to a new solution procedure. There is already a working procedure at hand, so why spend valuable time on trying to switch to a completely new algorithm even if the new algorithm may work better?

Being aware of all of these facts, I decided to write the book in order to aid both researchers and engineers in this regard. Strictly speaking, I would have never attempted to write such a book if the performances of the new procedures were not so wonderful. My decision came only after realizing that the new procedures would be foundational to a new generation of efficient and robust solvers for fluid flow problems.

With this in mind, I tried to design the book in such a way that a novice researcher can easily find only the specific material required in order to be able to write his own code, without much mathematical frustration. For the experienced researcher, I included various hints and tips so that the algorithms can be adapted to their existing codes quickly, without any problem or fear of failure. For this purpose, I have included FORTRAN codes for some of the procedures which may be called up directly. The only thing that an experienced researcher might need to do is to incorporate the boundary conditions describing the physical problem in a proper fashion and then call the routine, as described in the book.

I tried to do my best in smoothly introducing the new procedures and in designing the book so that it can further be used as a graduate textbook for engineers as well as mathematicians. With its present form, I hope that it can help in bridging the gap between the two disciplines and also act as a reference book for the experienced researchers.

I wish to acknowledge the late Prof. Bill Nicoll who introduced me to this fantastic field of science and to Prof. George D. Raithby for his enthusiastic encouragement that helped me progress in the field.

I would also like to express my special thanks to my wife Ümit, my son Erden, my daughter Eliz, my daughter in love Esin and my grand daughter Derin for their patience and loving encouragement during the course of writing the book. They deserved much more attention than I could devote to them during this time, but now I hope I will be able to redeem myself.

Contents

1 Introduction . . . 1
1.1 Scope of the Book . . . 1
1.2 Outline of the Book . . . 4

2 Preliminaries . . . 7
2.1 Quadratic Interpolation . . . 7
2.2 Approximations Using Lagrangian Polynomial . . . 8
2.3 Approximations Using Taylor Series . . . 10
2.4 General Elimination Technique for Linear Systems . . . 13
2.5 Solution Techniques for Special Linear Systems . . . 16

3 Governing Differential Equations . . . 19
3.1 Governing Equations . . . 19
3.2 Characteristics of the Governing Equations . . . 20
3.3 The Velocity-Pressure Coupling Problem . . . 21

4 Finite Difference Formulations . . . 25
4.1 Manipulation of the Momentum Equations . . . 25
4.2 Grid Arrangement for the Solution . . . 26
4.3 Profile Assumptions for the Discretizations . . . 28
4.4 Discretization of the Governing Equations . . . 31
4.4.1 Discretization of the *u*-Momentum Equation . . . 31
4.4.2 Discretization of the *v*-Momentum Equation . . . 35
4.4.3 Discretization of the Mass Conservation Equation . . . 38
4.5 A Discussion on the Profile Assumptions . . . 38

5 Preparations for Solution . . . 41
5.1 The Solution Region . . . 41
5.2 Boundary Conditions . . . 42

5.2.1 Types of Boundary Conditions 43
5.2.2 Application of Various Types of Boundary Conditions into the Momentum Equations. 45
5.2.3 Application of Boundary Conditions into the Coupled Equations 47
5.3 Incorporating Relaxation. 54

6 Assembling the Discretized Equations into a Block Matrix System. 57
6.1 The Numbering Scheme 57
6.2 Construction of the Block Matrix System 58
6.3 Disadvantages of the Block Matrix. 61

7 The Solution Procedure: Block Incomplete Decomposition 65
7.1 Properties and Advantages of the Block Matrix 66
7.2 General Incomplete Decomposition 68
7.3 An Incomplete Decomposition of the Block System (BIP) 71
7.4 The Block Solution Procedure. 77
7.5 Complete Solution of the Flow Field 78
7.6 A Family of Procedures: BIPEN, FICS-1, FICS-2 78
7.7 Storage Requirements and Complexity 82
7.7.1 Storage Requirements. 82
7.7.2 Complexity. 83
7.8 The Simplest Case (Simple Implicit Coupled Solution—SICS) 84
7.8.1 The Solution Algorithm (SICS). 85
7.8.2 Properties and Benefits of SICS. 87

8 Applications and Testing 89
8.1 Benchmark Fluid Flow Problems. 89
8.2 Testing Criteria 99
8.3 Performance Analysis and Comparisons 102
8.4 A Discussion of the Mechanism of the Procedures. 106
8.5 Comparison with the Segregated-Type Procedures 108
8.6 Convergence Characteristics and Performances of SICS and SIMPLER: A Relative Comparison 109

9 Special Cases. 115
9.1 Time-Dependent Problems 115
9.2 Stoke's Flow Equations 119
9.3 Turbulent Flows and Heat Transfer 119
9.4 Adaptation to Existing Codes 120
9.5 Three-Dimensional Problems. 120

10 Concluding Remarks 127

Appendix A: A Critical Survey of Literature—An Adventure into Perfection 129

Appendix B: Segregated Solution Procedures: SIMPLE and SIMPLER 141

Appendix C: FORTRAN Subroutines—BLOCKSOLFICS2 and BLOCKSOLSICS 159

References 167

Index 171

Nomenclature

A	Discretization coefficients
	Coefficient matrix
b	Right-hand-side coefficients
B	Right-hand-side vector
L, U	Strength matrices
D	Defect matrix
E	Under-relaxation factor
N	Total number of grid points
h, f, k, n	Strength vectors
a, x, b, s	
g, t, w, c	
e, q, d, y	Defect vectors
r, z, m	
m, n, l	Number of grid points in the x, y and z directions
N_{tot}	Total number of nodes
p	Pressure
u, v, w	Velocity components
x, y, z	Space coordinates
t	Time
δ, h, k	Mesh sizes
$\Delta x, \Delta y, \Delta z$	
f, Φ	General functions
θ, β	Weighting factor
K_x, K_y	Eddy diffusivity coefficients
κ	von Karman constant
ρ	Density

Subscripts

e, w, n, s, u, d	Cell faces
m, n, l	Number of nodes in x and y directions
S, W, P, E, N, D, U	Grid nodes
i	Indices for nodes
in	Indicates inflow
out	Indicates outflow

Superscripts

exact	Denotes exact value
C	Denotes continuity
p	Denotes pressure
u, v, w	Denotes velocity components
$*, c$	Denotes preliminary values
n	Iteration level

Chapter 1
Introduction

1.1 Scope of the Book

Fluid flow processes play a very important role in all aspects of human life. Nearly all industrial applications such as automotive design and machinery, factories, nuclear reactors, bridges, aircraft and ships, involve, one way or another, for their proper, economic and efficient functioning, fluid flow within or outside of each component. This may be air, water, gas, oil or even blood. At this point, understanding how the fluid flows in or around the objects and how much pressure is exerted on certain components plays a very important role in the efficient design of such equipment. Without this understanding, designing such equipments would never have been possible.

In early stages, such designs were accomplished by laboratory experiments. For this, first a prototype of the equipment had to be built and then tested in a suitable environment. The resulting deficiencies, then had to be corrected, and the equipment be redesigned and retested until the required performance was obtained. This, however, was a very expensive, time consuming and elaborate process.

Nowadays, thankfully, we have at our disposal a complete set of mathematical equations fully describing the physical properties of such fluid flows. These extend from simple boundary layer flows to the most complicated turbulent flows, be it two or three dimensional, time dependent or time independent. To these formulations are also added equations describing such extra processes as heat transfer or chemical processes inherent in the physical processes. Therefore, the design of such machinery is theoretically possible by predicting the behavior of the fluids involved, even before any physical prototype is built and tested. This can be accomplished simply by solving those equations under certain boundary conditions describing the physical situation under consideration.

With the availability of powerful computers nowadays, this seems to be a very attractive alternative. However, and this is a big however, there are certain important conditions that must be satisfied for this alternative to be realized.

© Springer International Publishing Switzerland 2016

Z. Mazhar, *Fully Implicit, Coupled Procedures in Computational Fluid Dynamics*, Fluid Mechanics and Its Applications 115, DOI 10.1007/978-3-319-29895-5_1

1. A numerical procedure, or a method is required in order to efficiently solve those nonlinear, coupled partial differential equations.
2. This procedure needs to be easy to understand and implement with a suitable computer code, even by a novice researcher who is not much of a professional in mathematics or numerical analysis.
3. The procedure must be robust, that is to say, it must work under tough conditions without much risk of possible divergence.
4. The method must be applicable to three-dimensional, as well as to two-dimensional problems.
5. The procedure must be applicable to steady as well as unsteady problems.
6. The procedure should require least computer storage, and it must be very fast.
7. Solution of other coupled equations describing varying physical properties and running parallel to the basic flow equations must also be possible.
8. Finally, there must be a well established, easy to understand reference guide for the user, in order to help with all aspects of the setup and implementation of the procedure.

In the heart of the complete set of partial differential equations lie a set of equations called momentum and mass conservation equations. The momentum equations mainly describe the behavior of the velocity components under pressure differences which may occur between various points of the flow field. These equations involve certain physical characteristics of the flow field and the fluid itself. These may either be constants or variables, or they may even be supplied by other phenomena such as heat transfer and turbulence.

To complement the momentum equations, there exists an additional equation called the mass conservation equation. This equation, although innocent looking and very simple both physically and mathematically, describes a strong connection between the velocity components of the flow. It does not, however, contain a term involving pressure so that it cannot be used as an explicit equation for pressure. So, since the momentum equations are described mainly for the velocities, there seems no direct way of calculating pressure. Indeed, this is a big problem and it is the main cause of the poor performance of most of the methods developed for solving these equations.

Various methods that have been developed and used extensively for the past fifty years or so are mostly based on a 'divide and conquer' strategy. This strategy provided a partial solution to the problem, but unfortunately its nature did not permit the development of high performance methods. Researchers, perhaps because they had no other choice, have spent much effort in revising these methods in an effort to handle the problem in a more efficient fashion. Since all of these efforts were circling around the basic old philosophy, not much advancement in computational efficiency was achieved. The problem was actually very well known, but not much effort has been spent toward answering the question of why these procedures were so slow and unreliable.

In recent years, the Author of the book has come up with reasonable answers to the above problem. It was only after the main reason of the poor performances was understood, that an alternative, well established mechanism could be developed to handle the problem in an efficient way. In contrast to the long-established, old-fashioned approaches of the past, the new procedures, thus constructed are based on a completely new way of looking at the problem, in which the discretized forms of the governing differential equations are solved in their primitive forms through a fully implicit, coupled mechanism. This mechanism enforces the full realization of any change in pressure or velocity components at all other points in the flow field even before the actual solution starts. The new strategy resulted in fast convergence rates and enhanced durability. Indications are that this new generation of procedures will be candidates for efficient use in the near future.

In order that a new procedure be widely utilized by researchers, the formulation of the mechanisms involved, together with any relevant details of the various aspects of the application, must be clearly understood. This, however, can be achieved by a proper guidance and detailed explanation of certain important aspects of the procedure. Without this proper guidance, the reader cannot easily adapt to the new procedure. Worse than this, the procedure might end up waiting in the dusty shelves of the literature. Hence the necessity for a book such as this.

Discretization of the governing equations, application of the involved boundary conditions, formation of a matrix system and solution of the resulting system of linear equations are the four basic steps in a solution procedure. Each of these steps involves various, finely tuned mathematical as well as computational details. An engineer without a fresh, broad and to-the-point background of such details, would never be able to form a proper solution routine. We recall the old saying that quotes 'The devil and perfection lay in the details'.

The book is designed exactly to address the above described problems. The first three steps in the solution process are given only with sufficient, but necessary details in order not to suffocate the reader with some irrelevant material. The last step, being the most important one, is given special attention. The underlying philosophy and the details of the construction of the procedures are given in an elaborate but easy to follow fashion. In this way the reader is smoothly introduced to a family of superbly efficient procedures whose performance and durability is far superior to any rivals.

What is presented in the book is not just a new single solution procedure, but actually a new mechanism which is based on a completely novel viewpoint, leading to a family of possible procedures. Hence, the mechanism paves the way for researchers to be able to construct their own varying procedures, depending on certain demands. In this regard, it is open for further ideas towards advancement.

From the possible family of procedures, one is so simple to construct and apply that it may take only a few hours for its adaptation to an existing code. Although it is a fully coupled procedure, it may be implemented like a point-by-point solver while requiring no extra storage for solution. This feature, combined with a high overall performance and speed, paves the way for its application to complicated grid structures in a straightforward manner. Furthermore, it is envisaged that it may even be utilized for some on-board, case oriented applications such as automatic shape changing mechanisms while functioning. This is a very important feature that could not even be imagined with a classical procedure.

The book is mainly designed to introduce the new family of procedures and the underlying philosophy. With its present form, it can also be regarded as a reference book for scientists, engineers and researchers working in the area of computational fluid dynamics. It is hoped that it will become an indispensable reference for such people who intend to write their own code with the aim of solving their problems using the most up-to-date computing methods in this field.

With this book as a reference, a novice, new to the field, will be able to understand all the necessary formulations and be empowered to implement the new procedures. The experienced reader, on the other hand, will be able to adapt the new procedures quickly to his own working code, with no hassle.

The book may also be used as a graduate textbook for engineers and mathematicians.

The Author feels that the book will fill a big gap in the literature in this field. In the meantime, it may also take the opportunity to demonstrate that Computational Fluid Dynamics is not just a tool for solving field problems, but also an enjoying area of engineering.

1.2 Outline of the Book

The book is designed in an orderly manner in order to explain step-by-step each and every aspect of the solution procedure in the simplest easy-to-follow form. Therefore, the user is urged, in the first reading, to follow the chapters of the book in sequence. Thereafter, it can be considered as a reference book in which the index will quickly direct one to a specific chapter or section that applies to the relevant application. The reader is assumed to have a suitable knowledge of the governing partial differential equations and the conditions under which they can describe a specific physical problem. Only a minimum knowledge of university level mathematics, such as calculus, differential equations, linear algebra and numerical analysis is assumed.

The underlying basic mathematical preliminaries are given in Chap. 2. In Chap. 3, the basic governing partial differential equations are given. These are the momentum and mass conservation equations, the solution of which is the core of all simulations of flow situations. The properties of, and the difficulties involved in their solution process are described.

Discretization of the governing equations is presented in detail, in Chap. 4. Suitable grid arrangements and profile assumptions are followed by a full description of the discretization process of all of the governing equations, including the mass conservation equation.

Some important preparatory operations such as boundary condition application and relaxation enforcement are discussed in Chap. 5.

Construction of a block matrix system by assembling the discretized forms of the governing equations is presented in Chap. 6. The difficulties involved in the solution of the block system are discussed.

The basic strategy used in the procedures, named the *F*ully *I*mplicit *C*oupled *S*olution (FICS), is given in Chap. 7, in which detailed derivations and formulations are described. Variations and alternate configurations of the procedures are discussed. A distinct, very simple mechanism, named *S*imple *I*mplicit *C*oupled *S*olution (SICS) is also introduced in a separate section.

In Chap. 8, various benchmark problems are introduced, and their solutions are given in order to aid the reader in the early stages of implementation and testing of the code. Convergence characteristics, solution costs and durability tests are presented in a comparative way. A section in this regard is devoted to the SICS procedure.

Chapter 9 involves discussions and recommendations for the application of the presented procedures for situations involving time dependency and the incorporation of other dependent variable equations into the momentum equations. Since the procedures can be extended for use in the solution of three dimensional problems as well, a special section is devoted to aid in this respect by formulating a three dimensional version of one of the procedures, namely SICS. A discussion on the adaptation of the block solution procedure for the readers already working with some other solution schemes is also included.

Throughout the main chapters, reference to any of the huge amount of literature is avoided intentionally. In this respect, the book is designed as a comprehensive and complete reference for the subject. However, a brief summary with a critical review of the history of the development of the procedures used in the literature in the previous decades is included in Appendix A for the interested reader. A brief overview of the construction of two basic such procedures is presented in Appendix B.

In complement, complete FORTRAN subroutines are provided for two of the block solution procedures, in Appendix C. The routines are ready to be used, with explanations of the input and output data.

Chapter 2
Preliminaries

The governing differential equations in fluid flow problems contain first and second order partial derivative terms. In order to obtain the solution of these equations, a process called 'discretization' must be applied onto these equations. This process involves the approximation of the derivative terms at each point of the solution region, by discrete functional values on and around these points. For this purpose, finite difference approximations of the derivative terms are required. These approximations can be obtained in various ways. One way is to use a quadratic interpolation function. Another way is to utilize Taylor series approximations around the point at which the approximation is made. This requires some knowledge of Taylor series. Utilizing quadratic interpolation is easier for the novice reader, but many are generally accustomed to using Taylor series. Taylor series has an advantage that the truncation error associated with the approximations is readily obtained. Here we will present both methods and leave the choice to the reader.

Moreover, in various stages of the block solutions, certain techniques are required for the solution of linear systems of equations which contain matrices of some special forms.

This chapter is devoted to the presentation of the mathematical preliminaries to be utilized for these purposes.

2.1 Quadratic Interpolation

Suppose we are given three points x_1, x_2 and x_3 in the x direction, and corresponding discrete functional values for a function $f(x)$ as $f_1 = f(x_1), f_2 = f(x_2)$ and $f_3 = f(x_3)$. If this is the only knowledge about the function $f(x)$, then the Fundamental Theorem of Linear Algebra can be used to pass a parabola of the form

$$f(x) = a + bx + cx^2 \tag{2.1}$$

© Springer International Publishing Switzerland 2016

Z. Mazhar, *Fully Implicit, Coupled Procedures in Computational Fluid Dynamics*,
Fluid Mechanics and Its Applications 115, DOI 10.1007/978-3-319-29895-5_2

from all of the three points (x_1,f_1), (x_2,f_2) and (x_3,f_3). We can obtain this parabola by substituting these values into Eq. (2.1) as

$$\begin{aligned} f_1 &= a + bx_1 + cx_1^2 \\ f_2 &= a + bx_2 + cx_2^2 \\ f_3 &= a + bx_3 + cx_3^2 \end{aligned} \tag{2.2}$$

These equations are linear in a, b and c, and thus can be easily solved. Solving these equations for a, b and c, replacing in Eq. (2.1) and rearranging terms give

$$f(x) = \frac{(x-x_2)(x-x_3)}{(x_1-x_2)(x_1-x_3)}f_1 + \frac{(x-x_1)(x-x_3)}{(x_2-x_1)(x_2-x_3)}f_2 + \frac{(x-x_1)(x-x_2)}{(x_3-x_1)(x_3-x_2)}f_3 \tag{2.3}$$

This equation is called the Lagrangian Polynomial. It is of a more suitable form for our purposes in the discretization of the derivatives in the governing differential equations.

Using this polynomial, we will derive expressions for the first and second order derivative terms $\frac{df}{dx}$ and $\frac{d^2f}{dx^2}$ of $f(x)$.

2.2 Approximations Using Lagrangian Polynomial

For notational purposes which will be clear later on in our formulations, we assume a grid arrangement with suitable notations as shown in Fig. 2.1.

The points W, P and E are the points x_1, x_2 and x_3 at which $f(x)$ is given as $f_1 = f_W, f_2 = f_P$ and $f_3 = f_E$, respectively. Points w and e are at midway between the points W and P, and P and E, respectively. Also note that $\delta_w = x_2 - x_1$ and $\delta_e = x_3 - x_2$.

Differentiating $f(x)$ in Eq. (2.3) with respect to x gives

$$\frac{df}{dx} = \frac{(x-x_3)+(x-x_2)}{(x_1-x_2)(x_1-x_3)}f_W + \frac{(x-x_3)+(x-x_1)}{(x_2-x_1)(x_2-x_3)}f_P + \frac{(x-x_2)+(x-x_1)}{(x_3-x_1)(x_3-x_2)}f_E \tag{2.4}$$

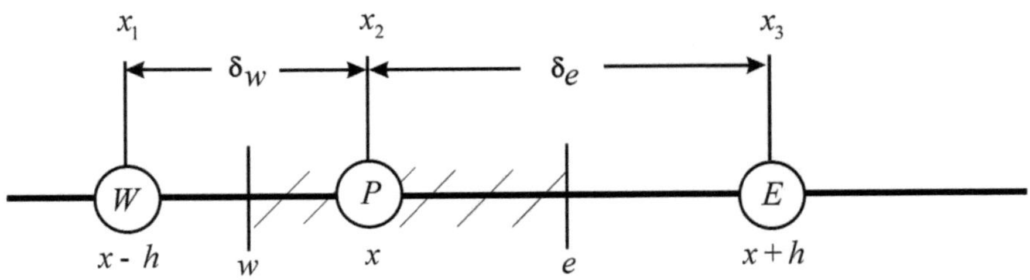

Fig. 2.1 Grid arrangement in the x direction

Now, the value of $\frac{\partial f}{\partial x}$ at the point $x_2 = P$ can be calculated by replacing x by x_2 in Eq. (2.4). This gives

$$\left.\frac{df}{dx}\right|_P = \frac{(x_2 - x_3) + (x_2 - x_2)}{(x_1 - x_2)(x_1 - x_3)} f_W + \frac{(x_2 - x_3) + (x_2 - x_1)}{(x_2 - x_1)(x_2 - x_3)} f_P + \frac{(x_2 - x_2) + (x_2 - x_1)}{(x_3 - x_1)(x_3 - x_2)} f_E \tag{2.5}$$

Replacing differences in x_i's with the corresponding δ_i's, gives

$$\left.\frac{df}{dx}\right|_P = -\frac{\delta_e}{\delta_w(\delta_e + \delta_w)} f_W + \frac{\delta_e - \delta_w}{\delta_e \delta_w} f_P + \frac{\delta_w}{\delta_e(\delta_w + \delta_e)} f_E \tag{2.6}$$

For the second order derivative $\frac{d^2 f}{dx^2}$, we differentiate $\frac{df}{dx}$ in Eq. (2.4) with respect to x which gives

$$\begin{aligned} \frac{d^2 f}{dx^2} = \frac{d}{dx}\left(\frac{df}{dx}\right) &= \frac{2}{(x_1 - x_2)(x_1 - x_3)} f_W + \frac{2}{(x_2 - x_1)(x_2 - x_3)} f_P \\ &+ \frac{2}{(x_3 - x_1)(x_3 - x_2)} f_E \end{aligned} \tag{2.7}$$

$\left.\frac{d^2 f}{dx^2}\right|_P$ can now be obtained as

$$\left.\frac{d^2 f}{dx^2}\right|_P = \frac{2}{\delta_w(\delta_e + \delta_w)} f_W - \frac{2}{\delta_e \delta_w} f_P + \frac{2}{\delta_e(\delta_w + \delta_e)} f_E \tag{2.8}$$

For a function $f(x)$ which is given continuously over the space x, the expressions given in Eqs. (2.6) and (2.8) for any three consecutive data points W, P and E, are only approximations to the corresponding derivatives. Therefore, their use in the governing differential equations leads only to approximations of the corresponding derivative terms. These approximations are exact only if the function $f(x)$ is a quadratic polynomial function, that is to say, a parabola of the form $f(x) = a + bx + cx^2$. Now, if the function is known to behave in a widely different form, the above approximations will not be appropriate because of the resulting poor approximations. In such a case, a numerical simulation of the unknown function $f(x)$ in which these approximations are used may not only give unreasonable results, but may well lead to unexpected divergence in the numerical solution as well. Therefore, if the behavior of the function is known, or even anticipated, a more suitable form of these approximations should be generated. The approximations will still be a linear combination of the three functional values, but with a slight alteration of the weighting factors. We will address this issue in Chap. 4 for the functions involved in the governing equations.

If we let $\delta_e = \delta_w = h$ in the above formulations, that is to say, if equidistant spacing between the x_i's is used, then Eqs. (2.6) and (2.8) become

$$\left.\frac{df}{dx}\right|_P = \frac{f_E - f_W}{2h} \tag{2.9}$$

and

$$\left.\frac{d^2f}{dx^2}\right|_P = \frac{f_W - 2f_P + f_E}{h^2} \tag{2.10}$$

respectively. These are the so called 'central difference' approximations for the derivative terms $\frac{df}{dx}$ and $\frac{d^2f}{dx^2}$ at a point P.

In the following chapters, for the sake of generality, we will use the non-equidistant spaced approximations as given in Eqs. (2.6) and (2.8).

2.3 Approximations Using Taylor Series

The Taylor's Theorem statest that under certain conditions, the value of a function at a point $x + h$, say $f(x + h)$, can be written in an infinite series form which contains the function and its derivatives at the point x as follows:

$$f(x+h) = f(x) + h\left.\frac{df}{dx}\right|_x + \left.\frac{h^2}{2!}\frac{d^2f}{dx^2}\right|_x + \left.\frac{h^3}{3!}\frac{d^3f}{dx^3}\right|_x + \cdots \tag{2.11}$$

If h is small enough and the higher order derivatives are bounded, we can write an approximation to $f(x + h)$ by dropping the terms containing derivatives of order three and higher, as

$$f(x+h) \approx f(x) + h\left.\frac{df}{dx}\right|_x + \left.\frac{h^2}{2!}\frac{d^2f}{dx^2}\right|_x \tag{2.12}$$

If we apply this formula by considering Fig. 2.1, we can write

$$f_E \approx f_P + \delta_e\left.\frac{df}{dx}\right|_P + \left.\frac{\delta_e^2}{2!}\frac{d^2f}{dx^2}\right|_P \tag{2.13}$$

and

$$f_W \approx f_P - \delta_w\left.\frac{df}{dx}\right|_P + \left.\frac{\delta_w^2}{2!}\frac{d^2f}{dx^2}\right|_P \tag{2.14}$$

To obtain an approximation for $\frac{df}{dx}\big|_P$, we can now manipulate Eqs. (2.13) and (2.14) by eliminating the terms $\frac{d^2f}{dx^2}\Big|_P$. For this, we multiply Eq. (2.13) by $\frac{2}{\delta_e^2}$ and Eq. (2.14) by $-\frac{2}{\delta_w^2}$ and add the two resulting expressions side by side. We get

$$\frac{2}{\delta_e^2}f_E - \frac{2}{\delta_w^2}f_W \approx \left[\frac{2}{\delta_e^2} - \frac{2}{\delta_w^2}\right]f_P + \left[\frac{2}{\delta_e} + \frac{2}{\delta_w}\right]\frac{df}{dx}\bigg|_P \tag{2.15}$$

Rearranging Eq. (2.15), we can write

$$\frac{\delta_e + \delta_w}{\delta_e\delta_w}\frac{df}{dx}\bigg|_P \approx -\frac{1}{\delta_w^2}f_W + \frac{\delta_e^2 - \delta_w^2}{\delta_e^2\delta_w^2}f_P + \frac{1}{\delta_e^2}f_E \tag{2.16}$$

from which,

$$\frac{df}{dx}\bigg|_P \approx -\frac{\delta_e}{\delta_w(\delta_e + \delta_w)}f_W + \frac{\delta_e - \delta_w}{\delta_e\delta_w}f_P + \frac{\delta_w}{\delta_e(\delta_w + \delta_e)}f_E \tag{2.17}$$

This is the same as the expression derived in Eq. (2.6).

To obtain the approximation for $\frac{d^2f}{dx^2}\Big|_P$, this time we must eliminate the terms $\frac{df}{dx}\big|_P$. For this, we multiply Eq. (2.13) by $\frac{1}{\delta_e}$ and Eq. (2.14) by $\frac{1}{\delta_w}$ and add the two expressions side by side. We get

$$\frac{1}{\delta_e}f_E + \frac{1}{\delta_w}f_W \approx \left[\frac{1}{\delta_e} + \frac{1}{\delta_w}\right]f_P + \left[\frac{\delta_e}{2} + \frac{\delta_w}{2}\right]\frac{d^2f}{dx^2}\bigg|_P \tag{2.18}$$

Rearranging Eq. (2.18) gives

$$\left[\frac{\delta_e + \delta_w}{2}\right]\frac{d^2f}{dx^2}\bigg|_P \approx \frac{1}{\delta_w}f_W - \left[\frac{\delta_e + \delta_w}{\delta_e\delta_w}\right]f_P + \frac{1}{\delta_e}f_E \tag{2.19}$$

from which

$$\frac{d^2f}{dx^2}\bigg|_P \approx \frac{2}{\delta_w(\delta_e + \delta_w)}f_W - \frac{2}{\delta_e\delta_w}f_P + \frac{2}{\delta_e(\delta_w + \delta_e)}f_E \tag{2.20}$$

which is the same as the expression derived in Eq. (2.8).

For the purposes that will be clear later on, it is beneficial to consider approximations for the above derivatives in the y direction as well. This will aid in our

Fig. 2.2 Grid arrangement in the y direction

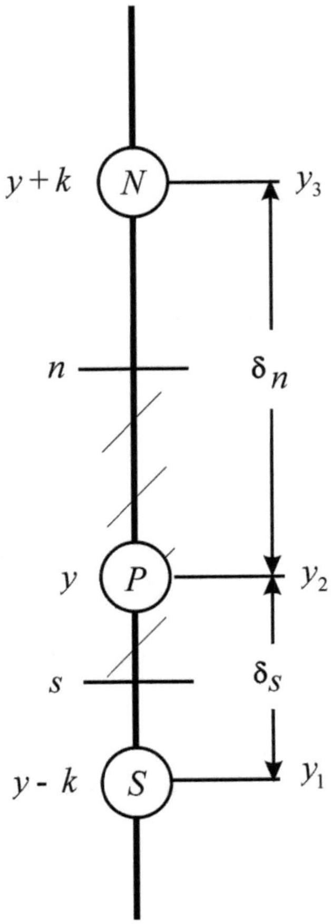

notational convention to be used for the partial derivatives in the governing differential equations of our interest.

This time we consider a grid arrangemet in the y direction as shown in Fig. 2.2.

Suppose that we are given a function $f(y)$. Now, following similar derivations as shown above, we can write approximations for the derivatives $\frac{df}{dy}$ and $\frac{d^2f}{dx^2}$ at the point P as follows:

$$\left.\frac{df}{dy}\right|_P \approx -\frac{\delta_n}{\delta_s(\delta_n+\delta_s)}f_S + \frac{\delta_n-\delta_s}{\delta_n\delta_s}f_P + \frac{\delta_s}{\delta_n(\delta_s+\delta_n)}f_N \tag{2.21}$$

$$\left.\frac{d^2f}{dx^2}\right|_P \approx \frac{2}{\delta_s(\delta_n+\delta_s)}f_S - \frac{2}{\delta_n\delta_s}f_P + \frac{2}{\delta_n(\delta_s+\delta_n)}f_N \tag{2.22}$$

2.4 General Elimination Technique for Linear Systems

A linear system of equations with N equations and N unknowns can be written as

$$\begin{aligned} a_{11}x_1 + a_{12}x_2 + \cdots + a_{1N}x_N &= b_1 \\ a_{21}x_1 + a_{22}x_2 + \cdots + a_{2N}x_N + &= b_2 \\ &\vdots \\ a_{N1}x_1 + a_{N2}x_2 + \cdots + a_{NN}x_N - &= b_N \end{aligned} \tag{2.23}$$

in which a_{ij}, $i = 1, N$ and $j = 1, N$ are constant coefficients and x_i, $i = 1, N$ are the unknown variables. A solution to this system is the set of x_i values which satisfy all of the equations simultaneously. This system is written in a compact form called a matrix equation as follows:

$$\begin{bmatrix} a_{11} & a_{12} & \cdots & a_{1N} \\ a_{21} & a_{22} & \cdots & a_{2N} \\ \vdots & \vdots & \ddots & \vdots \\ a_{N1} & a_{N2} & \cdots & a_{NN} \end{bmatrix} \begin{bmatrix} x_1 \\ x_2 \\ \vdots \\ x_N \end{bmatrix} = \begin{bmatrix} b_1 \\ b_2 \\ \vdots \\ b_N \end{bmatrix} \tag{2.24}$$

The matrix

$$A = \begin{bmatrix} a_{11} & a_{12} & \cdots & a_{1N} \\ a_{21} & a_{22} & \cdots & a_{2N} \\ \vdots & \vdots & \ddots & \vdots \\ a_{N1} & a_{N2} & \cdots & a_{NN} \end{bmatrix}$$

is called the **coefficient matrix** and

$$A_{aug} = \begin{bmatrix} a_{11} & a_{12} & \cdots & a_{1N} & b_1 \\ a_{21} & a_{22} & \cdots & a_{2N} & b_2 \\ \vdots & \vdots & \ddots & \vdots & \vdots \\ a_{N1} & a_{N2} & \cdots & a_{NN} & b_N \end{bmatrix}$$

is called the **augmented matrix** of the system.

One of the basic strategies for solving the above system directly is to apply a series of operations, called **elementary row operations** onto the rows of the augmented matrix and update it to an equivalent form that is easier to solve. The following operations are used for this purpose, each of which does not have any effect on the solution set of the equations:

1. Multiply all entries in a row by a nonzero constant
2. Replace one row by adding to it another row, coefficient by coefficient
3. Interchange the places of two rows

The aim in the first step of the solution process is to eliminate all of the nonzero coefficients in the lower triangular part of the augmented matrix which corresponds to the matrix of the original system. To eliminate the coefficient a_{21} of the augmented matrix, i.e., to make it zero, we multiply row 1 by $-\dfrac{a_{21}}{a_{11}}$

$$A_{aug'} = \begin{bmatrix} a_{11} & a_{12} & \cdots & a_{1N} & b_1 \\ 0 & a'_{22} & \cdots & a'_{2N} & b'_2 \\ \vdots & \vdots & \ddots & \vdots & \vdots \\ a_{N1} & a_{N2} & \cdots & a_{NN} & b_N \end{bmatrix}$$

and add it to row 2. We get

We note that all the coefficients in row 2 have changed now. Then we proceed to eliminate a_{31}, a_{41}

until a_{N1}, in the same way. At the end of this process, the augmented matrix becomes

$$A_{aug'} = \begin{bmatrix} a_{11} & a_{12} & \cdots & a_{1N} & b_1 \\ 0 & a'_{22} & \cdots & a'_{2N} & b'_2 \\ \vdots & \vdots & \ddots & \vdots & \vdots \\ 0 & a'_{N2} & \cdots & a'_{NN} & b'_N \end{bmatrix}$$

We repeat the elimination process for the coefficients a_{32}, a_{42} until a_{N2}. The whole process is continued until the coefficient $a_{N,N-1}$ is eliminated.

The complete process is named as the **forward elimination**, after which we have an **upper triangular matrix**, in which all of the coefficients in the lower triangular part are zeros.

$$A_{aug'} = \begin{bmatrix} a_{11} & a_{12} & \cdots & a_{1N} & b_1 \\ 0 & a'_{22} & \cdots & a'_{2N} & b'_2 \\ \vdots & \vdots & \ddots & \vdots & \vdots \\ 0 & 0 & \cdots & a'_{NN} & b'_N \end{bmatrix}$$

Now the second step in the solution process is to obtain the solution to the original system by applying a **back substitution process.** For this, we solve x_N from the Nth equation, x_{N-1} from equation $(N-1)$ etc., until x_1.

An algorithm for the complete solution process can now be written as follows:
Forward elimination:

```
for k=1,N-1
   for i=k+1,N
      t=a(i,k)/a(k,k)
     a(i,k)=0
     for j=k+1,N
       a(i,j)=a(i,j)-a(k,j)*t
     end j
     b(i)=b(i)-b(k)*t
   end i
end k
```

Backward substitution:

```
x(N)=b(N)/a(N,N)
for i=N-1,1,-1
   sum=0
   for j=i+1,N
      sum=sum+a(i,j)*x(j)
   end j
   x(i)=(b(i)-sum)/a(i,i)
end i
```

The process described is named as the **Gauss Elimination method** in the literature.

We note that the third elementary row operation has not been utilized in this process. If, however, any diagonal coefficient a_{ii} becomes zero at any stage of the forward elimination process, row i must be interchanged by another row having a nonzero coefficient on the main diagonal. Otherwise, a division by zero will occur while the back substitution process is performed. Luckily enough, as we will see later on, the matrix systems that are involved in incompressible fluid flow solutions do not pose this problem, if of course, the systems are formed properly. Hence this issue will not be elaborated any further.

2.5 Solution Techniques for Special Linear Systems

Within the block solution procedures described later, various linear sub-systems whose matrices are of a certain special form need to be solved. Therefore, a brief review of the techniques used to solve such systems is appropriate.

Type 1: Lower triangular systems with two off-diagonal vectors
Suppose we have a system of linear equations given as

$$\begin{bmatrix} c_1 & & & & & & & \\ b_2 & c_2 & & & & & & \\ & b_3 & c_3 & & & & & \\ & & \cdot & \cdot & & & & \\ a_{m+1} & & & \cdot & \cdot & & & \\ & \cdot & & & \cdot & \cdot & & \\ & & \cdot & & & \cdot & \cdot & \\ & & & a_N & & & b_3 & c_N \end{bmatrix} \begin{bmatrix} x_1 \\ x_2 \\ x_3 \\ \cdot \\ \cdot \\ \cdot \\ \cdot \\ x_N \end{bmatrix} = \begin{bmatrix} r_1 \\ r_2 \\ r_3 \\ \cdot \\ \cdot \\ \cdot \\ \cdot \\ r_N \end{bmatrix}$$

This matrix contains only two vectors in the lower triangular part and the upper triangular part is completely zero. One of the vectors is just below the main diagonal vector, and the other is far below by m columns. All other coefficients in the lower diagonal part are zeros. Here it is possible to store the complete matrix by utilizing only three vectors in computer's memory.

The system can be solved easily by a forward substitution sweep using the following algorithm:

```
x(1)=r(1)/c(1)
for i=2,m
   x(i)=(r(i)-b(i)*x(i-1))/c(i)
end i
for i=m+1,N
   x(i)=(r(i)-a(i)*x(i-m)-b(i)*x(i-1))/c(i)
end i
```

Type 2: Upper triangular systems with two off-diagonal vectors
In this case we consider the following linear system of equations:

$$\begin{bmatrix} c_1 & d_1 & & & e_1 & & & \\ & c_2 & d_2 & & & \cdot & & \\ & & c_3 & d_3 & & & \cdot & \\ & & & \cdot & \cdot & & & e_{N-m} \\ & & & & \cdot & \cdot & & \\ & & & & & \cdot & \cdot & \\ & & & & & & \cdot & d_{N-1} \\ & & & & & & & c_N \end{bmatrix} \begin{bmatrix} x_1 \\ x_2 \\ x_3 \\ \cdot \\ \cdot \\ \cdot \\ \cdot \\ x_N \end{bmatrix} = \begin{bmatrix} r_1 \\ r_2 \\ r_3 \\ \cdot \\ \cdot \\ \cdot \\ \cdot \\ r_N \end{bmatrix}$$

Here we have only three vectors in the upper triangular part. All other coefficients, including those at the lower triangular part are zeros. For this system, a backward substitution sweep is applied, using the following algorithm:

```
x(N)=r(N)/c(N)
for i=N-1,N-m+1,-1
   x(i)=(r(i)-d(i)*x(i+1))/c(i)
end i
for i=N-m,1,-1
   x(i)=(r(i)-d(i)*x(i+1)-e(i)*x(i+m))/c(i)
end i
```

Type 3: A tri-diagonal system
Suppose we are given the following system which contains a diagonal vector and two off-diagonal vectors just below and above this diagonal vector as follows:

$$\begin{bmatrix} b_1 & c_1 & & & & \\ a_2 & b_2 & c_2 & & & \\ & a_3 & b_3 & c_3 & & \\ & & \cdot & \cdot & \cdot & \\ & & & \cdot & \cdot & \cdot \\ & & & & \cdot & \cdot & \cdot \\ & & & & & \cdot & \cdot & c_N \\ & & & & & & a_N & b_3 \end{bmatrix} \begin{bmatrix} x_1 \\ x_2 \\ x_3 \\ \cdot \\ \cdot \\ \cdot \\ \cdot \\ x_N \end{bmatrix} = \begin{bmatrix} r_1 \\ r_2 \\ r_3 \\ \cdot \\ \cdot \\ \cdot \\ \cdot \\ r_N \end{bmatrix}$$

This system can be solved by a forward sweep to eliminate the coefficients a_i, followed by a backward sweep to obtain the solution. An algorithm similar to the one used in the Gaussian Elimination method can be used here. Application of the usual algorithm as given above however, would change the values of the coefficients of the vectors. But for reasons that will be clear later on, we do not want to spoil the original values of the coefficients of the three vectors. We can achieve this by not actually updating the a_i's and utilizing an auxiliary vector t for keeping the updated versions of the b_i's. The c_i's do not change, so they are kept untouched. The following algorithm performs the forward elimination and backward substitution processes.

```
t(1)=b(1)
for i=2,N
   q=a(i)/t(i-1)
   t(i)=b(i)-c(i-1)*q
   r(i)=r(i)-r(i-1)*q
end i
x(N)=r(N)/t(N)
for i=N-1,1,-1
   x(i)=(r(i)-c(i)*x(i+1))/t(i)
end i
```

Type 4: Lower triangular systems with three off-diagonal vectors
Suppose we are given the system

$$\begin{bmatrix} c_1 & & & & & & & \\ b_2 & c_2 & & & & & & \\ & b_3 & c_3 & & & & & \\ e_m & & \cdot & \cdot & & & & \\ a_{m+1} & e_{m+1} & & \cdot & \cdot & & & \\ & \cdot & \cdot & & \cdot & \cdot & & \\ & & \cdot & \cdot & & \cdot & \cdot & \\ & & & a_N & e_N & & b_3 & c_N \end{bmatrix} \begin{bmatrix} x_1 \\ x_2 \\ x_3 \\ \cdot \\ \cdot \\ \cdot \\ \cdot \\ x_N \end{bmatrix} = \begin{bmatrix} r_1 \\ r_2 \\ r_3 \\ \cdot \\ \cdot \\ \cdot \\ \cdot \\ r_N \end{bmatrix}$$

In this case the matrix is similar to that given in Type 1, but it contains an additional vector just above the lower far-diagonal vector. This system can be solved by a forward sweep using the following algorithm:

```
x(1)=r(1)/c(1)
for i=2,m-1
   x(i)=(r(i)-b(i)*x(i-1))/c(i)
end i
x(m)=(r(m)-e(m)*x(i-m+1)-b(i)*x(i-1))/c(i)
for i=m+1,N
   x(i)=(r(i)-a(i)*x(i-m)-e(i)*x(i-m+1)-b(i)*x(i-1))/c(i)
end i
```

Type 5: Upper triangular systems with three off-diagonal vectors
The system to be solved is given by

$$\begin{bmatrix} c_1 & d_1 & & y_1 & e_1 & & & \\ & c_2 & d_2 & & \cdot & \cdot & & \\ & & c_3 & d_3 & & \cdot & \cdot & \\ & & & \cdot & \cdot & & \cdot & e_{N-m} \\ & & & & \cdot & \cdot & & y_{N-m+1} \\ & & & & & \cdot & \cdot & \\ & & & & & & \cdot & d_{N-1} \\ & & & & & & & c_N \end{bmatrix} \begin{bmatrix} x_1 \\ x_2 \\ x_3 \\ \cdot \\ \cdot \\ \cdot \\ \cdot \\ x_N \end{bmatrix} = \begin{bmatrix} r_1 \\ r_2 \\ r_3 \\ \cdot \\ \cdot \\ \cdot \\ \cdot \\ r_N \end{bmatrix}$$

In this case, the coefficient matrix is similar to that in Type 2, but it contains an additional vector just below the far-diagonal vector. The algorithm given below uses a backward substitution sweep to solve this system:

```
x(N)=r(N)/c(N)
for i=N-1,N-m+2,-1
   x(i)=(r(i)-d(i)*x(i+1))/c(i)
end i
i=N-m+1
x(i)=(r(i)-d(i)*x(i+1)-y(i)*x(i+m-1))/c(i)
for i=N-m,1,-1
   x(i)=(r(i)-d(i)*x(i+1)-y(i)*x(i+m-1)-e(i)*x(i+m))/c(i)
end i
```

Chapter 3
Governing Differential Equations

The basic governing differential equations for fluid flow consist of a set of momentum equations, and a mass conservation or continuity equation.

In order to be able to devise a useful solution strategy for these equations, it is necessary to have a broad and precise understanding of the special features of these equations. For sure, a correct diagnosis to the difficulties involved can bring a more efficient remedy for the solution of these equations.

With the above in mind, we first introduce the basic governing equations for an incompressible fluid flow. The formulations are for two-dimensional problems since a two-dimensional representation is more feasible for a better understanding of the special techniques used in the solution procedures. This is followed by a discussion of the special features of the equations and the difficulties involved regarding their solutions. This discussion will help in realizing an insight into the choice of the solution procedures presented in the remaining chapters.

3.1 Governing Equations

The partial differential equations governing the two-dimensional, mean motion of a steady flow of an incompressible fluid may be written as

$$\frac{\partial}{\partial x}\left(K_x\frac{\partial u}{\partial x}\right)+\frac{\partial}{\partial y}\left(K_y\frac{\partial u}{\partial y}\right)-\frac{\partial}{\partial x}(\rho uu)-\frac{\partial}{\partial y}(\rho uv)-\frac{\partial p}{\partial x}=0 \tag{3.1}$$

(u momentum equation)

$$\frac{\partial}{\partial x}\left(K_x\frac{\partial v}{\partial x}\right)+\frac{\partial}{\partial y}\left(K_y\frac{\partial v}{\partial y}\right)-\frac{\partial}{\partial x}(\rho uv)-\frac{\partial}{\partial y}(\rho vv)-\frac{\partial p}{\partial y}=0 \tag{3.2}$$

(v momentum equation)

© Springer International Publishing Switzerland 2016

Z. Mazhar, *Fully Implicit, Coupled Procedures in Computational Fluid Dynamics*, Fluid Mechanics and Its Applications 115, DOI 10.1007/978-3-319-29895-5_3

$$\frac{\partial u}{\partial x} + \frac{\partial v}{\partial y} = 0 \text{ (mass conservation equation)} \tag{3.3}$$

In the above equations x is the horizontal space coordinate, y is the vertical space coordinate, u and v are the mean velocity components in the x and y directions, respectively. K_x and K_y are the eddy diffusivity coefficients for momentum in the x and y directions, respectively. ρ and p are density and pressure.

For a particle in a flow field to be able to move from one point to another, there must exist a pressure difference between the two points, represented by the pressure differential terms on the right sides of the two momentum equations. If the pressure differences are all zero and there are no other external forces acting on the field, the fluid remains at standstill. The higher the pressure difference between the points, the higher the velocity in that direction. Therefore, it is obvious that there exists a very important relationship between the velocity field and the pressure field.

The mass conservation equation completing the set, includes only the velocity field components and mainly describes that mass should be preserved in all of the cells of the flow field and also globally over the solution region. Mass preservation means that any amount of fluid flowing into a fluid cell must flow out of the cell as well.

3.2 Characteristics of the Governing Equations

To understand the mechanism of the solution procedures discussed in the following chapters, and why such a mechanism is used, the characteristics of these equations should be examined more closely. First, we note the followings:

1. Both of the momentum equations are nonlinear in u and v due to the products in the terms containing first order derivatives. Therefore, any numerical approximation scheme to be utilized must incorporate a linearization process to deal with the nonlinearities. This can be accomplished by some kind of 'overall iterations'.
2. Due to some high pressure differences, abrupt changes in the velocity field may occur from one overall iteration to another. The effect of these changes must be fully transformed and distributed to the whole of the flow field. If this cannot be done efficiently, these changes may need to be artificially suppressed by certain amounts. Otherwise the process may collapse.
3. All the unknown functions u, v and p appear in the two momentum equations. Due to the presence of the second order derivative terms for u and v, these equations seem to stand mainly for the functions u and v.
4. Any changes in pressure, represented by the pressure difference terms in the momentum equations cause changes in the velocity components as well.
5. The most innocent-looking equation is the mass conservation equation. It contains only the unknowns u and v. It represents a very strong relationship between

u and v through the first order derivatives. In addition to obeying this dictation, the velocity components must also satisfy the momentum equations.

6. Despite its importance, however, the mass conservation equation does not contain the pressure function. Therefore, it is clear that virtually there exists no explicit equation describing the behavior of the pressure field, in contrast to its importance in the development of the velocity field.
7. From the momentum equations point of view, it is noteworthy that not the pressure itself, but actually the pressure differences affect the velocity field. Therefore, it is not important what the pressure magnitudes are. This is called the *relativeness* of the pressure values. The deciding factor for the velocity field are the pressure changes between two neighboring points in the flow field.

3.3 The Velocity-Pressure Coupling Problem

It is in fact a reality that, the mass conservation equation plays a very important role in the behavior of the velocity field, but it has no direct effect upon the pressure field. On the contrary, it is merely a rude constraint on the velocity field invoked by the pressure differences. This constitutes a one-way relationship, since the constraint does not contain pressure.

The absence of the pressure function in the mass conservation equation causes a very important problem. In fact, despite that there exists a very strong coupling between the velocity and pressure fields, there is no explicit equation for pressure. In the literature, this is called the ***velocity-pressure coupling*** problem. Owing to its importance and the difficulties involved in its cure, some researchers describe this problem as the ***'bottle-neck'*** of the solution of fluid flow problems.

However, as we will demonstrate later on, the above observation is actually an illusion which misled researchers and diverted their attention to 'creating' an equation for pressure.

For decades, researchers have proposed various methods for dealing with this problem. In fact, any numerical scheme which does not pose a remedy for this problem cannot be used efficiently for the solution of any fluid flow problem. Such a scheme must posses a strong capability for the treatment of this problem. Here we will not attempt to go into the details of such previous schemes, but rather give a brief summary of the common drawbacks of these procedures.

Considering that there exists no explicit equation for pressure, researchers tend to produce (or rather 'invent') an equation for it. A brief review of two of the widely used ones is given in Appendix B. Such 'fictitious' equations were first used to correct the velocity field. Then, they were used to update the pressure field. But the various procedures for generating such equations were cumbersome, complicated and confusing, not even mentioning the physical destruction of the unrealistic assumptions they involved.

The equations, thus generated for pressure are of elliptic type (or Poisson-type) whose solution is a more challenging problem as compared to the solution of the momentum equations with an assumed pressure field. This is because the involved boundary conditions for these 'pressure' equations are of Neumann-type, that is to say, at all of the boundaries, pressure differences must be set to zero. Pressure can only be set to a fixed value at only one point within the solution domain.

The numerical solution of the Poisson-type equations, furthermore, poses important convergence problems. Firstly the methods for solving such elliptic equations are very slow or even divergent if no strong relaxation is used. The relaxation parameters need to be finely tuned since convergence intervals of the relaxation parameters used are very narrow. Realizing these deficiencies, some more efficient solution methods were even generated for these equations.

Luckily however, some strategies involved in these methods have been used in the development of the type of procedures presented in this book.

There is a well-known, basic, but very important property of the flow of an incompressible fluid, which was unfortunately overlooked in the past decades. In fact, it was this negligence that prevented the development of more efficient procedures for the solution of field problems.

It is well known that a fluid particle may only move from one point to another if there exists a pressure difference between the two points. The movement is initiated and dictated by the pressure derivative terms in the momentum equations, and then guided by the diffusion and convection terms. While this movement is underway, the particle's velocity components must obey the mass conservation constraint. Thus, the effects of the pressure derivative terms in the momentum equations and those of the velocity derivative terms in the mass conservation equation are equivalently important. These effects are related to the fluid flow itself, but not to the physical properties which are governed by the diffusion and convection terms. Therefore, in order that a solution procedure be more efficient, instead of concentrating on the convection and diffusion terms, as is done with the segregated procedures, it must concentrate on the correct imposition of the pressure changes as well as the mass conservation constraint in all stages of the solution process.

This observation asserts that the priority must be given to the fluid flow properties rather than the physical properties of the fluid itself. To put it more precisely, the effect of the pressure terms, in parallel to the mass conservation principle, must be regarded as a constraint, perhaps to be better named as the 'pressure conservation constraint'. The importance of this issue will be more appreciated while we introduce the new solution strategy in Chap. 7.

On the other hand, it is clear by examining the governing equations that a pressure change at a point in the solution domain does not only affect the velocities at nearby points, but also the velocities at all other points in the solution domain. Therefore, any such changes must immediately be appreciated by the mechanism of the numerical procedure and transported to all of the points in an efficient fashion. In a point by point or line by line solution procedure, this appreciation cannot be realized immediately at all points.

The main problem in the segregated type procedures is the dissatisfaction of the mass and momentum conservation constraints while the solution is progressing. In such procedures, the constraints are only satisfied when and if complete convergence is reached. This 'if', however, is of utmost importance which deserves much attention.

The flow development in an incompressible fluid is characterized by sudden and perhaps 'unexpected' changes in either the velocity field or the pressure field. If the procedure used lacks the ability to quickly and efficiently distribute the effect of these changes through the solution region, the mass and momentum constraints will be violated at the corresponding control volumes. This violation, especially of the mass conservation constraint, even at one single control volume, may cause a deadly numerical crash.

Although very well known, an efficient solution to this crucial problem was never pursued. The problem was 'wiped out under the rug' by resorting to an 'easy' solution. This 'corner cutting' solution was to 'suppress' the changes by some heavy under relaxation process, in an effort to avoid divergence. Such a suppression, although it may succeed, inevitably and dramatically slows down the solution process. Physically, this sums up to a rude numerical intervention into the natural behavior of the flow field.

Then, why should we interfere with the natural development of the flow field? Isn't there some way of preserving its natural behavior without resorting to physically unrealistic assumptions or interventions? The answers to the above questions are: No, we must not interfere much, and yes, there is some way of preservation.

With the introduction of the so called 'Block Implicit Decomposition' in the last decade, many such problems have been overcome.

The general principle in this strategy is to let the pressure differences act freely onto the velocity field, but in the meantime, take necessary precautions to distribute the resulting effects correctly and efficiently to the whole field.

Some properties and advantages of these procedures which are presented in Chap. 7, may be summarized as follows:

1. The governing equations are taken 'as is'. After proper discretization of each equation, the resulting discretized equations involving velocity components and pressure, are accepted as they are given. That is to say, no artificial pressure equation is derived and no unrealistic assumptions are made. The main advantage here is that since no assumptions are made, convergence is not suppressed but rather encouraged.
2. The numerous relaxation parameters and the need for their tuning is avoided. Some relaxation, of course, is necessary for the nonlinearities and for speeding up the convergence in the solution process. However, the strong implicitness realized with such procedures so broadened the range of these few parameters that, with acceptance of some slightly slower convergence, any or no relaxation would suffice. Relaxation need only be applied for further speeding up convergence.
3. The derivations and application of the solution mechanism are extremely simple.

Chapter 4
Finite Difference Formulations

This chapter is devoted to introduce, in full, the discretization process for the governing equations, utilizing some enhanced versions of the approximations derived in Chap. 2. The chapter is designed as follows: In Sect. 4.1, the momentum equations given in Eqs. (3.1) and (3.2) are manipulated in order to impose the mass conservation condition in Eq. (3.3). The grid arrangement and certain profile assumptions suitable for discretization are given in Sects. 4.2 and 4.3. The discretization of the three governing equations are given in detail in Sect. 4.4. A discussion of the special profile assumptions for the velocity components is included in Sect. 4.5.

4.1 Manipulation of the Momentum Equations

Before continuing any further, the following manipulation of the momentum equations is appropriate:

First, we note that the term $\frac{\partial}{\partial x}(\rho uu)$ in Eq. (3.1) can be written as

$$\frac{\partial}{\partial x}(\rho uu) = \rho\frac{\partial}{\partial x}\left(u^2\right) = 2\rho u\frac{\partial u}{\partial x}, \tag{4.1}$$

upon differentiation of the term $\frac{\partial}{\partial x}(u^2)$ with respect to x.

On the other hand, the term $\frac{\partial}{\partial y}(\rho uv)$ in Eq. (3.1) can be written as

$$\frac{\partial}{\partial y}(\rho uv) = \rho\frac{\partial}{\partial y}(uv) = \rho\left[u\frac{\partial v}{\partial y} + v\frac{\partial u}{\partial y}\right] \tag{4.2}$$

© Springer International Publishing Switzerland 2016

Z. Mazhar, *Fully Implicit, Coupled Procedures in Computational Fluid Dynamics*, Fluid Mechanics and Its Applications 115, DOI 10.1007/978-3-319-29895-5_4

Adding Eqs. (4.1) and (4.2) gives

$$\begin{aligned}\frac{\partial}{\partial x}(\rho uu)+\frac{\partial}{\partial y}(\rho uv) &= 2\rho u\frac{\partial u}{\partial x}+\rho\left[u\frac{\partial v}{\partial y}+v\frac{\partial u}{\partial y}\right]\\ &= \rho\left[2u\frac{\partial u}{\partial x}+u\frac{\partial v}{\partial y}+v\frac{\partial u}{\partial y}\right]\end{aligned}\tag{4.3}$$

Now noting from Eq. (3.3) that $\frac{\partial v}{\partial y}=-\frac{\partial u}{\partial x}$ and replacing $\frac{\partial v}{\partial y}$ by $-\frac{\partial u}{\partial x}$ in Eq. (4.3) we get

$$\begin{aligned}\frac{\partial}{\partial x}(\rho uu)+\frac{\partial}{\partial y}(\rho uv) &= \rho\left[2u\frac{\partial u}{\partial x}-u\frac{\partial u}{\partial x}+v\frac{\partial u}{\partial y}\right]\\ &= \rho\left[u\frac{\partial u}{\partial x}+v\frac{\partial u}{\partial y}\right]\end{aligned}\tag{4.4}$$

Then Eq. (3.1) becomes

$$\frac{\partial}{\partial x}\left(K_x\frac{\partial u}{\partial x}\right)+\frac{\partial}{\partial y}\left(K_y\frac{\partial u}{\partial y}\right)-\rho\left[u\frac{\partial u}{\partial x}+v\frac{\partial u}{\partial y}\right]-\frac{\partial p}{\partial x}=0\tag{4.5}$$

As an exercise, the user may now be able to manipulate the corresponding terms in Eq. (3.2) to get

$$\frac{\partial}{\partial x}\left(K_x\frac{\partial v}{\partial x}\right)+\frac{\partial}{\partial y}\left(K_y\frac{\partial v}{\partial y}\right)-\rho\left[u\frac{\partial v}{\partial x}+v\frac{\partial v}{\partial y}\right]-\frac{\partial p}{\partial x}=0\tag{4.6}$$

Although the mass conservation condition has already been imposed on the momentum equations, Eqs. (4.5) and (4.6) are going to be used as the main governing equations along with the mass conservation Eq. (3.3). It should have been clear in the reader's mind that this imposition may help in the robustness of the procedures used, and in fact it does.

The main aim here is to present an efficient solution procedure of Eqs. (4.5) and (4.6) under the constraint of the mass conservation equation given in Eq. (3.3).

4.2 Grid Arrangement for the Solution

The first step in any numerical solution procedure is to decide on a suitable grid system which extends over the whole of the solution domain. The solution domain is 'discretized' into a finite number of grid points placed within the domain of interest. Since the mass conservation constraint states on its own that the amount of fluid entering a fluid cell must be equal to the exiting fluid, it is reasonable and advantageous to choose the grid structure in such a way that this condition is easily and exactly fulfilled at each cell. If in addition, necessary precautions are taken at

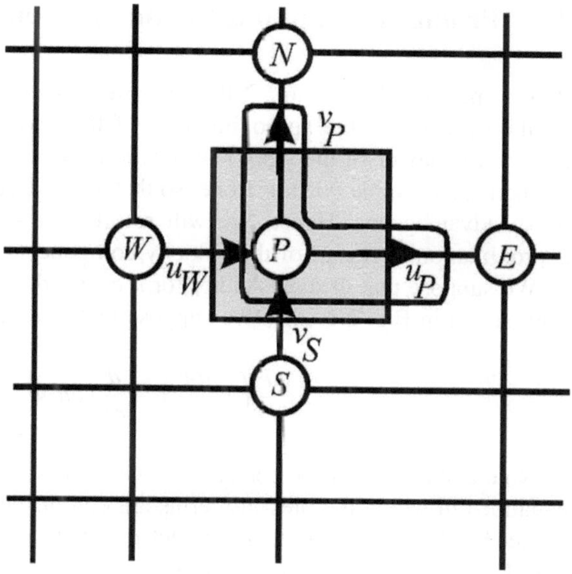

Fig. 4.1 Typical control volume and storage configuration

the boundaries of the flow region, then the constraint will also be satisfied globally. This can be accomplished by ensuring that the entering and exiting fluid velocity components are placed at the correct positions.

The second step is to locate the velocity and pressure variables so that they can be coupled more tightly.

With these in mind, the following 'control volume' and 'staggered grid' approach widely used in the literature is quite adequate. Typical of such a control volume and storage configuration is shown in Fig. 4.1.

In Fig. 4.1, we note that the velocity components u and v and the pressure variable p are located at different grid points in the domain, but u and v exactly at the faces of the control volume. This may look a little bit of unusual for the novice reader, but the advantages and simplicity of such a placement will be clear very soon.

The following convention is then used: the value of u stored at a point numbered by (P) is the 'actual' u value at the midway between the points designated by (P) and (E); the value of v velocity stored at (P) is the 'actual' v value at the midway between the points designated by (P) and (N). This, naturally, necessitates the use of different grid configurations for the discretization of the two momentum equations and for the mass conservation equation.

4.3 Profile Assumptions for the Discretizations

It was mentioned in Chap. 2 that certain problems may be encountered in the solution process if the approximations of the derivatives do not incorporate the expected behavior of the unknown function. In our case, the expected behavior is not perhaps a simple quadratic one, so that a direct quadratic interpolation may not be so advantageous. Hence, we will resort to a slight revision of the quadratic approximation in regard of the velocity components as follows.

We suppose that Φ stands either for u or v and let this function prevail over the grid shown in Fig. 2.1. Also we suppose that Φ satisfies the differential equation

$$\frac{d}{dx}\left(K\frac{d\Phi}{dx}\right) - \frac{d}{dx}(\rho u^* \Phi) = 0 \tag{4.7}$$

Note that this equation represents a one dimensional form of the two momentum equations in which the pressure terms are eliminated.

Let K, u^* and ρ be constants, and let $x_2 = 0$ and $x_3 = L$ in Fig. 2.1. Suppose also that Φ satisfies the boundary conditions

$$\Phi(0) = \Phi_P \text{ and } \Phi(L) = \Phi_E.$$

A slight refreshment of college differential equations knowledge can give the exact analytical solution of this problem over the interval $(0, L) = (0, \delta_e)$ as

$$\frac{\Phi - \Phi_P}{\Phi_E - \Phi_P} = \frac{e^{\frac{\rho u^*}{K}x} - 1}{e^{\frac{\rho u^*}{K}L} - 1} \tag{4.8}$$

Now an expression for Φ_e can be written by noting that $L = \delta_e$ and the coordinate of the point e is δ_e, as

$$\frac{\Phi_e - \Phi_P}{\Phi_E - \Phi_P} = \frac{e^{\frac{\rho u^* \delta_e}{2K}} - 1}{e^{\frac{\rho u^* \delta_e}{K}} - 1} \tag{4.9}$$

from which

$$\Phi_e = \Phi_P + (\Phi_E - \Phi_P)\frac{e^{\frac{\rho u^* \delta_e}{2K}} - 1}{e^{\frac{\rho u^* \delta_e}{K}} - 1} \tag{4.10}$$

or

$$\Phi_e = (1 - \theta)\Phi_P + \theta\Phi_E \tag{4.11}$$

in which

$$\theta = \frac{e^{\frac{\rho u^* \delta_e}{2K}} - 1}{e^{\frac{\rho u^* \delta_e}{K}} - 1} \tag{4.12}$$

Letting $r_e = \frac{\rho u_e^* \delta_e}{K}$ and $\theta = \frac{1}{2} - \alpha_e$, Eq. (4.11) can be written as

$$\Phi_e = (\frac{1}{2} + \alpha_e)\Phi_P + (\frac{1}{2} - \alpha_e)\Phi_E \tag{4.13}$$

in which

$$\alpha_e = \frac{1}{2} - \frac{e^{r_e/2} - 1}{e^{r_e} - 1} \tag{4.14}$$

A similar expression for Φ_w can be written as

$$\Phi_w = (\frac{1}{2} + \alpha_w)\Phi_W + (\frac{1}{2} - \alpha_w)\Phi_P \tag{4.15}$$

in which

$$\alpha_w = \frac{1}{2} - \frac{e^{r_w/2} - 1}{e^{r_w} - 1} \tag{4.16}$$

where

$$r_w = \frac{\rho u_w^* \delta_w}{K} \tag{4.17}$$

We note that instead of a straight averaging of the functional values around the points e and w, a variable averaging is used depending on the values of α_e and α_w in Eqs. 4.13 and 4.15, respectively.

Using these interpolations, we now work out approximations for the differential terms $\frac{d\Phi}{dx}\Big|_e$ and $\frac{d\Phi}{dx}\Big|_w$:

First we write Eq. 4.8 as

$$\Phi = \Phi_P + (\Phi_E - \Phi_P)\frac{e^{\frac{\rho u^*}{K}x} - 1}{e^{\frac{\rho u^*}{K}L} - 1} \tag{4.18}$$

Differentiating Φ with respect to x we get

$$\frac{d\Phi}{dx} = (\Phi_E - \Phi_P)\frac{\rho u^*}{K}\frac{e^{\frac{\rho u^*}{K}x}}{e^{\frac{\rho u^*}{K}L} - 1} \tag{4.19}$$

Now $\frac{d\Phi}{dx}\Big|_e$ can be written by letting $x = \delta_e/2$ and $L = \delta_e$, as

$$\left.\frac{d\Phi}{dx}\right|_e = (\Phi_E - \Phi_P)\frac{\rho u_e^*}{K}\frac{e^{\frac{\rho u_e^*}{2K}\delta_e}}{e^{\frac{\rho u^*}{K}\delta_e} - 1} \tag{4.20}$$

or

$$\left.\frac{d\Phi}{dx}\right|_e = \beta_e \frac{\Phi_E - \Phi_P}{\delta_e} \tag{4.21}$$

in which

$$\beta_e = r_e \frac{e^{r_e/2}}{e^{r_e} - 1} \tag{4.22}$$

The coefficients α's and β's are called *weighting factors*. When $r = 0$, the calculation of these coefficients will give an overflow on a computer due to the 0/0 condition. However, as $r \rightarrow 0$, $\alpha \rightarrow 0$ and $\beta \rightarrow 1$. Moreover, since the calculation of the exponentials in these factors is machine-expensive, the following approximations have been used in the literature, which do not include the above drawback (see Appendix A):

$$\alpha = \frac{r^2}{10 + 2r^2} \tag{4.23}$$

and

$$\beta = \frac{1 + 0.005r^2}{1 + 0.05r^2} \tag{4.24}$$

The exact and approximate values for the weighting factors α and β are shown in Fig. 4.2.

Similar approximations for $\left.\frac{d\Phi}{dx}\right|_w$ can be be written as

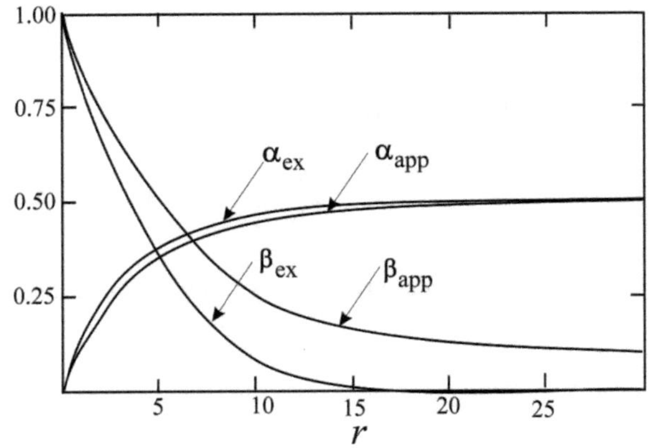

Fig. 4.2 Weighting factors α and β

$$\left.\frac{d\Phi}{dx}\right|_{w} = \beta_w \frac{\Phi_P - \Phi_W}{\delta_w} \tag{4.25}$$

It is left to the reader to write corresponding approximations in the y direction as well.

Equations (4.13), (4.15), (4.21) and (4.25) together with the weighting factors as described in Eqs. (4.23) and (4.24) will be utilized in the next section for the discretization of the governing differential equations.

4.4 Discretization of the Governing Equations

This section is fully devoted to the details of the discretization of the governing Eqs. (4.5), (4.6) and (3.3) for all points in the solution domain. The formulations are given in detail.

4.4.1 Discretization of the u*-Momentum Equation*

To begin with the discretization of the u-momentum Eq. (4.5), we consider the 'control volume' around an arbitrary point P in the solution domain as shown in Fig. 4.3.

Here we note that the value of the unknown u at the point P, denoted by u_P, has four neighboring functional values, denoted by u_E, u_W, u_N and u_S at the east, west,

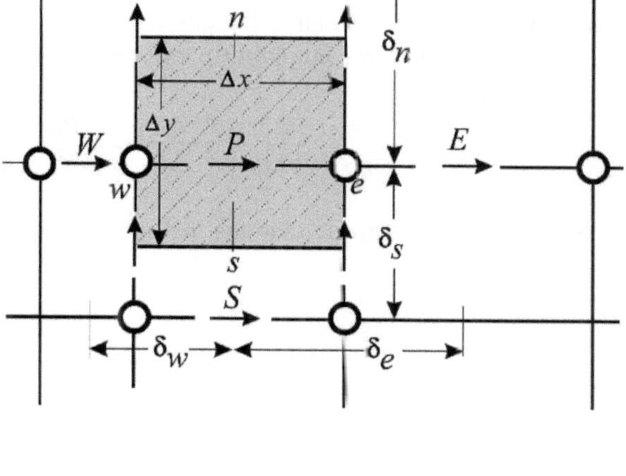

Fig. 4.3 Control volume for u velocity

north and south sides, respectively. These points are placed midway between the two neighboring grid points, or the pressure points, in both directions.

Now we use the approximations derived in Eqs. (4.13), (4.15), (4.21) and (4.25) to approximate each term in the u momentum equation as follows:

$$\frac{\partial}{\partial x}\left(K_x\frac{\partial u}{\partial x}\right)_P = K_x\frac{\partial}{\partial x}\left(\frac{\partial u}{\partial x}\bigg|_P\right) \approx \frac{K_x}{\Delta x}\left[\frac{\partial u}{\partial x}\bigg|_e - \frac{\partial u}{\partial x}\bigg|_w\right]$$
$$= \frac{K_x}{\Delta x}\left[\beta_e\frac{u_E - u_P}{\delta_e} - \beta_w\frac{u_P - u_W}{\delta_w}\right] \tag{4.26}$$

$$\frac{\partial}{\partial y}\left(K_y\frac{\partial u}{\partial y}\right)_P = K_y\frac{\partial}{\partial y}\left(\frac{\partial u}{\partial y}\bigg|_P\right) \approx \frac{K_x}{\Delta y}\left[\frac{\partial u}{\partial y}\bigg|_n - \frac{\partial u}{\partial y}\bigg|_s\right]$$
$$= \frac{K_y}{\Delta y}\left[\beta_n\frac{u_N - u_P}{\delta_n} - \beta_s\frac{u_P - u_S}{\delta_s}\right] \tag{4.27}$$

$$\left[u\frac{\partial u}{\partial x} + v\frac{\partial u}{\partial y}\right]_P \approx u_P^*\frac{\partial u}{\partial x}\bigg|_P + v_P^*\frac{\partial u}{\partial y}\bigg|_P$$
$$= u_P^*\left[\frac{u_e - u_w}{\Delta x}\right] + v_P^*\left[\frac{v_n - v_s}{\Delta y}\right]$$
$$= \frac{u_P^*}{\Delta x}\left[\left\{(\tfrac{1}{2}+\alpha_e)u_P + (\tfrac{1}{2}-\alpha_e)u_E\right\} - \left\{(\tfrac{1}{2}+\alpha_w)u_W + (\tfrac{1}{2}-\alpha_w)u_P\right\}\right]$$
$$- \frac{v_P^*}{\Delta y}\left[\left\{(\tfrac{1}{2}+\alpha_n)u_P + (\tfrac{1}{2}-\alpha_n)u_N\right\} - \left\{(\tfrac{1}{2}+\alpha_s)u_S + (\tfrac{1}{2}-\alpha_s)u_P\right\}\right] \tag{4.28}$$

A discussion on the relationship between the weighting applied in the above discretizations, with those that were present in the approximation of the derivatives in Eqs. (2.6) and (2.8) is provided in Sect. 4.5.

Here we note that u_P^* and v_P^* are the values of u_P and v_P that were assumed to be 'constant' at the discretization level. Actually, while the solution progresses, they will be values from the previous iteration level, or values at the initial level. The starred values become equal to the un-starred ones at the overall convergence level.

Continuing, the term involving pressure can be approximated as

$$\frac{\partial p}{\partial x}\bigg|_P \approx \frac{p_e - p_w}{\Delta x} = \frac{p_E - p_P}{\frac{1}{2}(\delta_e + \delta_w)} \tag{4.29}$$

Substituting the approximations in Eqs. (4.26)–(4.29) into Eq. (4.5) we get

$$
\begin{aligned}
&\frac{K_x}{\Delta x}\left[\beta_e\frac{u_E-u_P}{\delta_e}-\beta_w\frac{u_P-u_W}{\delta_e}\right]+\frac{K_y}{\Delta y}\left[\beta_n\frac{u_N-u_P}{\delta_n}-\beta_s\frac{u_P-u_S}{\delta_s}\right]\\
&\quad-\frac{\rho u_P^*}{\Delta x}\left[\left\{(\tfrac{1}{2}+\alpha_e)u_P+(\tfrac{1}{2}-\alpha_e)u_E\right\}-\left\{(\tfrac{1}{2}+\alpha_w)u_W+(\tfrac{1}{2}-\alpha_w)u_P\right\}\right]\\
&\quad-\frac{\rho v_P^*}{\Delta y}\left[\left\{(\tfrac{1}{2}+\alpha_n)u_P+(\tfrac{1}{2}-\alpha_n)u_N\right\}-\left\{(\tfrac{1}{2}+\alpha_s)u_S+(\tfrac{1}{2}-\alpha_s)u_P\right\}\right]\\
&\quad+\frac{p_E-p_P}{\Delta x}=0
\end{aligned}
\tag{4.30}
$$

Now multiplying all sides in Eq. (4.30) by the product $\Delta x\Delta y$ and collecting terms we get

$$
\begin{aligned}
&\left[K_y\Delta x\frac{\beta_s}{\delta_s}+\rho v_P^*\Delta x(\tfrac{1}{2}+\alpha_s)\right]u_S\\
&\quad+\left[K_x\Delta y\frac{\beta_w}{\delta_w}+\rho u_P^*\Delta y(\tfrac{1}{2}+\alpha_w)\right]u_W\\
&\quad-\left[\begin{array}{l}K_x\Delta y\left(\frac{\beta_e}{\delta_e}+\frac{\beta_w}{\delta_w}\right)+K_y\Delta x\left(\frac{\beta_n}{\delta_n}+\frac{\beta_s}{\delta_s}\right)\\ \qquad +{}_P^{*}\Delta y\left\{(\tfrac{1}{2}+\alpha_e)-(\tfrac{1}{2}-\alpha_w)\right\}+\rho v_P^*\left\{(\tfrac{1}{2}+\alpha_n)-(\tfrac{1}{2}-\alpha_s)\right\}\end{array}\right]u_P\\
&\quad+\left[K_x\frac{\beta_e}{\delta_e}-\rho u_P^*(\tfrac{1}{2}-\alpha_e)\right]u_E\\
&\quad+\left[K_y\Delta x\frac{\beta_n}{\delta_n}-\rho v_P^*\Delta x(\tfrac{1}{2}-\alpha_n)\right]u_N\\
&\quad+[\Delta y]p_P-[\Delta y]p_E=0
\end{aligned}
\tag{4.31}
$$

Equation (4.31) can now be written as

$$
A_S^u u_S+A_W^u u_W+A_P^u u_P+A_E^u u_E+A_N^u u_N+A_P^{pu}p_P+A_E^{pu}p_E=b_P^u \tag{4.32}
$$

in which

$$
A_S^u=\Delta x\left[K_y\frac{\beta_s}{\delta_s}+\rho v_P^*(\tfrac{1}{2}+\alpha_s)\right] \tag{4.33a}
$$

$$
A_W^u=\Delta y\left[K_y\frac{\beta_s}{\delta_s}+\rho u_P^*(\tfrac{1}{2}+\alpha_w)\right] \tag{4.33b}
$$

$$
A_E^u=\Delta y\left[K_x\frac{\beta_e}{\delta_e}-\rho u_P^*(\tfrac{1}{2}-\alpha_e)\right] \tag{4.33c}
$$

$$A_N^u = \Delta x\left[K_y\frac{\beta_n}{\delta_n} - \rho v_P^*(\tfrac{1}{2} - \alpha_n)\right] \tag{4.33d}$$

$$A_E^{pu} = -\Delta y \tag{4.33e}$$

$$A_P^{pu} = \Delta y \tag{4.33f}$$

$$A_P^u = -(A_S^u + A_W^u + A_E^u + A_N^u) \tag{4.33g}$$

and

$$b_P^u = 0 \tag{4.33h}$$

The validity of Eq. 4.33g is easy and it is left to the reader as an exercise.

There is an important issue to be discussed here, before going any further. We note from Sect. 4.3 that the factor r, defined by $r = \rho u\delta/K$, always carries the sign of u. On the other hand, α, when defined exactly by Eqs. (4.14) and (4.16) in terms of r, carries the sign of r too. Therefore α carries the sign of u as well. However, if the approximation for α as described in Eqs. (4.23) and (4.25) are used, as is done in Eqs. (4.33a–4.33d), α will still be positive in a case when u is negative. This discrepancy will lead to physically unrealistic results at the points where the velocities u^* and v^* are negative. To correct this deficiency, Eqs. (4.33a–4.33d) must be updated as follows:

$$\begin{aligned} A_S^u &= \Delta x\left[K_y\frac{\beta_s}{\delta_s} + \tfrac{1}{2}\rho v_P^* + |\rho v_P^*|\alpha_s\right] \\ &= \Delta x\left[\frac{(1+0.005r_s^2)}{(1+0.05r_s^2)}\frac{K_y}{\delta_s} + \frac{1^*}{2P} + \frac{1}{2}|\rho v_P^*|\frac{r_s^2}{5+r_s^2}\right] \end{aligned} \tag{4.33a$'$}$$

$$\begin{aligned} A_W^u &= \Delta y\left[K_y\frac{\beta_w}{\delta_w} + \tfrac{1}{2}\rho u_P^* + |\rho u_P^*|\alpha_w\right] \\ &= \Delta y\left[\frac{(1+0.005r_w^2)}{(1+0.05r_w^2)}\frac{K_y}{\delta_w} + \tfrac{1}{2}\rho u_P^* + \tfrac{1}{2}|\rho u_P^*|\frac{r_w^2}{5+r_w^2}\right] \end{aligned} \tag{4.33b$'$}$$

$$\begin{aligned} A_E^u &= \Delta y\left[K_x\frac{\beta_e}{\delta_e} - \tfrac{1}{2}\rho u_P^* + |\rho u_P^*|\alpha_e\right] \\ &= \Delta y\left[\frac{(1+0.005r_e^2)}{(1+0.05r_e^2)}\frac{K_x}{\delta_e} - \tfrac{1}{2}\rho u_P^* + \tfrac{1}{2}|\rho u_P^*|\frac{r_e^2}{5+r_e^2}\right] \end{aligned} \tag{4.33c$'$}$$

$$\begin{aligned} A_N^u &= \Delta x\left[K_y\frac{\beta_n}{\delta_n} - \tfrac{1}{2}\rho u_P^* + |\rho v_P^*|\alpha_n\right] \\ &= \Delta x\left[\frac{(1+0.005r_n^2)}{(1+0.05r_n^2)}\frac{K_y}{\delta_n} - \tfrac{1}{2}\rho u_P^* + \tfrac{1}{2}|\rho v_P^*|\frac{r_n^2}{5+r_n^2}\right] \end{aligned} \tag{4.33d$'$}$$

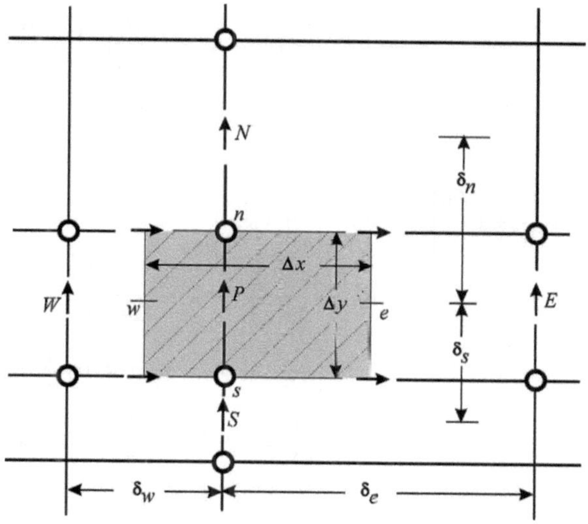

Fig. 4.4 Control volume for v velocity

4.4.2 Discretization of the v-Momentum Equation

The discretization of the v momentum equation is much the same as for the u momentum equation. The only difference is that this time we consider the control volume for the v variable as shown in Fig. 4.4.

Here it is important to note that the values of the mesh differences δ_e, δ_w, δ_n, δ_s, Δx and Δy are not the same as in the discretized forms of the u momentum equation due to the different control volumes used. Therefore the reader must be careful while calculating the coefficients for each momentum equation.

We consider the 'control volume' around an arbitrary point P in the solution domain and note that the neighboring functional values of v_P at the east, west, north and south sides of the point P are v_E, v_W, v_N and v_S, respectively. As in the u momentum equation these points are placed midway between the two neighboring pressure points on the north and south sides of the control volume.

With these in mind, the approximations derived in Eqs. (4.13), (4.15), (4.21) and (4.25) are now used to approximate each term in the v-momentum Eq. (4.6) as follows:

$$\frac{\partial}{\partial x}\left(K_x\frac{\partial v}{\partial x}\right)_P = K_x\frac{\partial}{\partial x}\left(\left.\frac{\partial v}{\partial x}\right|_P\right) \approx \frac{K_x}{\Delta x}\left[\left.\frac{\partial v}{\partial x}\right|_e - \left.\frac{\partial v}{\partial x}\right|_w\right]$$

$$= \frac{K_x}{\Delta x}\left[\beta_e\frac{v_E - v_P}{\delta_e} - \beta_w\frac{v_P - v_W}{\delta_e}\right] \tag{4.34}$$

$$
\begin{aligned}
\frac{\partial}{\partial y}\left(K_y \frac{\partial v}{\partial y}\right)_P &= K_y \frac{\partial}{\partial y}\left(\left.\frac{\partial v}{\partial y}\right|_P\right) \approx \frac{K_x}{\Delta y}\left[\left.\frac{\partial v}{\partial y}\right|_n - \left.\frac{\partial v}{\partial y}\right|_s\right] \\
&= \frac{K_y}{\Delta y}\left[\beta_n \frac{v_N - v_P}{\delta_n} - \beta_s \frac{v_P - v_S}{\delta_s}\right]
\end{aligned} \tag{4.35}
$$

$$
\begin{aligned}
\left[u\frac{\partial v}{\partial x} + v\frac{\partial v}{\partial y}\right]_P &\approx u_P^* \left.\frac{\partial v}{\partial x}\right|_P + v_P^* \left.\frac{\partial v}{\partial y}\right|_P \\
&= u_P^*\left[\frac{v_e - v_w}{\Delta x}\right] + v_P^*\left[\frac{u_n - u_s}{\Delta y}\right] \\
&= \frac{u_P^*}{\Delta x}\left[\left\{(\tfrac{1}{2} + \alpha_e)v_P + (\tfrac{1}{2} - \alpha_e)v_E\right\} - \left\{(\tfrac{1}{2} + \alpha_w)v_W + (\tfrac{1}{2} - \alpha_w)v_P\right\}\right] \\
&\quad - \frac{v_P^*}{\Delta y}\left[\left\{(\tfrac{1}{2} + \alpha_n)v_P + (\tfrac{1}{2} - \alpha_n)v_N\right\} - \left\{(\tfrac{1}{2} + \alpha_s)v_S + (\tfrac{1}{2} - \alpha_s)v_P\right\}\right]
\end{aligned} \tag{4.36}
$$

The term involving pressure is approximated as

$$
\left.\frac{\partial p}{\partial y}\right|_P \approx \frac{p_N - p_P}{\Delta y} \tag{4.37}
$$

Substituting the approximations in Eqs. (4.34)–(4.37) into Eq. (4.6), multiplying by $\Delta x \Delta y$ and rearranging we get

$$
\begin{aligned}
&\Delta x\left[K_y \frac{\beta_s}{\delta_s} + \rho v_P^*(\tfrac{1}{2} + \alpha_s)\right] v_S + \Delta y\left[K_x \frac{\beta_w}{\delta_w} + \rho u_P^*(\tfrac{1}{2} + \alpha_w)\right] v_W \\
&- \left[\begin{aligned} &K_x\left(\tfrac{\beta_e}{\delta_e} + \tfrac{\beta_w}{\delta_w}\right) + K_y \Delta x\left(\tfrac{\beta_n}{\delta_n} + \tfrac{\beta_s}{\delta_s}\right) \\ &+ \rho u_P^* \Delta y\{(\tfrac{1}{2} + \alpha_e) - (\tfrac{1}{2} - \alpha_w)\} + {}_P^*\Delta x\{(\tfrac{1}{2} + \alpha_n) - (\tfrac{1}{2} - \alpha_s)\} \end{aligned}\right] v_P \\
&+ \Delta y\left[K_x \frac{\beta_e}{\delta_e} - \rho u_P^*(\tfrac{1}{2} - \alpha_e)\right] v_E \\
&+ \left[K_y \frac{\beta_n}{\delta_n} - \rho v_P^*(\tfrac{1}{2} - \alpha_n)\right] v_N \\
&+ [\Delta x]\, p_P - [\Delta x]\, p_N = 0
\end{aligned} \tag{4.38}
$$

which can be cast into

$$
A_S^v v_S + A_W^v v_W + A_P^v v_P + A_E^v v_E + A_N^v v_N + A_P^{pv} p_P + A_N^{pv} p_N = b_P^v \tag{4.39}
$$

in which

$$A_S^u = \Delta x \left[K_y \frac{\beta_s}{\delta_s} + \rho v_P^* (\frac{1}{2} + \alpha_s) \right] \tag{4.40a}$$

$$A_W^u = \Delta y \left[K_y \frac{\beta_s}{\delta_s} + \rho u_P^* (\frac{1}{2} + \alpha_w) \right] \tag{4.40b}$$

$$A_E^u = \Delta y \left[K_x \frac{\beta_e}{\delta_e} - \rho u_P^* (\frac{1}{2} - \alpha_e) \right] \tag{4.40c}$$

$$A_N^u = \Delta x \left[K_y \frac{\beta_n}{\delta_n} - \rho v_P^* (\frac{1}{2} - \alpha_n) \right] \tag{4.40d}$$

$$A_N^{pv} = -\Delta x \tag{4.40e}$$

$$A_P^{pv} = \Delta x \tag{4.40f}$$

$$A_P^v = -(A_S^v + A_W^v + A_E^v + A_N^v) \tag{4.40g}$$

and

$$b_P^v = 0 \tag{4.40h}$$

Here also we note that if the approximations for the weighting factors α and β are used as described in Sect. 4.3, then Eqs. (4.40a–4.40d) must be updated similarly as discussed in the previous subsection. This is straightforward and it is left to the reader.

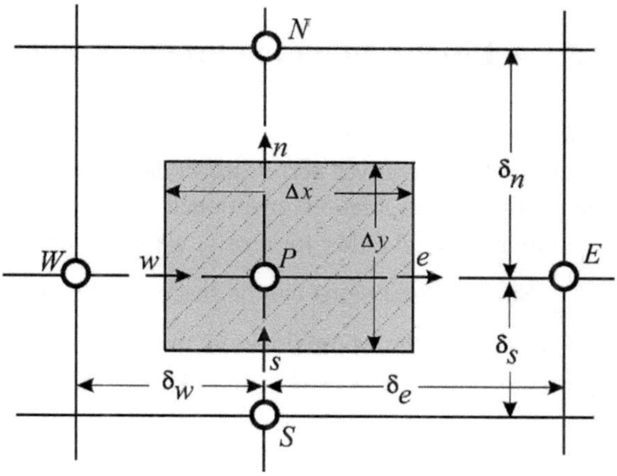

Fig. 4.5 Control volume for mass conservation

4.4.3 Discretization of the Mass Conservation Equation

For the discretization of the mass conservation Eq. (3.3) we use the mass control volume shown in Fig. 4.5.

We get

$$\frac{u_e - u_w}{\Delta x} + \frac{v_n - v_s}{\Delta y} = 0 \tag{4.41}$$

Since this equation is going to be used together with the discretized forms of the momentum equations, it is now suitable to divert to the same notation for node numbering. Noting that the actual velocity values lie exactly at the four faces of the control volume for mass conservation, we can write $u_e = u_P$, $u_w = u_W$, $v_n = v_P$ and $v_s = v_S$. Therefore Eq. (4.41) can be written as

$$\frac{u_P - u_W}{\Delta x} + \frac{v_P - v_S}{\Delta y} = 0 \tag{4.42}$$

Multiplying Eq. (4.42) by $\Delta x \Delta y$ and rearranging, we get

$$[-\Delta y]u_W + [\Delta y]u_P + [-\Delta x]v_S + [\Delta x]v_P = 0 \tag{4.43}$$

which can be written in the form

$$A_W^{Cu}u_W + A_P^{Cu}u_P + A_S^{Cv}v_S + A_P^{Cv}v_P = b_P^p \tag{4.44}$$

in which $A_W^{Cu} = -\Delta y$, $A_W^{Cu} = \Delta y$, $A_S^{Cv} = -\Delta x$, $A_W^{Cv} = \Delta x$ and $b_P^p = 0$.

We note here that b_P^p is zero, but we will demonstrate later that it may become nonzero for a point neighboring a boundary, upon implementing certain boundary conditions regarding the velocities at the boundary.

4.5 A Discussion on the Profile Assumptions

Before closing the chapter, a discussion of the approximations for the first and second order derivatives performed in the above sections and their relationship with the quadratic interpolations mentioned in Chap. 2 is appropriate.

Letting Φ stand either for u or v, we will work out approximations for $\left.\frac{\partial \Phi}{\partial x}\right|_P$ and $\left.\frac{\partial^2 \Phi}{\partial x^2}\right|_P$. Similar discussions can be made for the derivatives in the y direction.

Considering the approximation made in Eq. (4.28) regarding $\left.\frac{\partial \Phi}{\partial x}\right|_P$ with the approximations in Eqs. (4.13) and (4.15) in Sect. 4.3, we can write

$$\begin{aligned}
\left.\frac{\partial \Phi}{\partial x}\right|_P &\approx \frac{\Phi_e - \Phi_w}{\Delta x} = \frac{\Phi_e - \Phi_w}{(\delta_e + \delta_w)/2} \\
&= \frac{2}{\delta_e + \delta_w}\left[(\tfrac{1}{2} + \alpha_e)\Phi_P + (\tfrac{1}{2} - \alpha_e)\Phi_E - (\tfrac{1}{2} + \alpha_w)\Phi_W - (\tfrac{1}{2} - \alpha_w)\Phi_P\right] \\
&= \frac{2}{\delta_e + \delta_w}\left[-(\tfrac{1}{2} + \alpha_w)\Phi_W + (\alpha_e + \alpha_w)\Phi_P + (\tfrac{1}{2} - \alpha_e)\Phi_E\right] \\
&= -\frac{(1 + 2\alpha_w)}{\delta_e + \delta_w}\Phi_W + \frac{2(\alpha_e + \alpha_w)}{\delta_e + \delta_w}\Phi_P + \frac{(1 - 2\alpha_e)}{\delta_e + \delta_w}\Phi_E
\end{aligned} \tag{4.45}$$

Now replacing α_e and α_w in Eq. (4.45) by $\alpha_e = \frac{\delta_e - \delta_w}{2\delta_e}$ and $\alpha_w = \frac{\delta_e - \delta_w}{2\delta_w}$ gives the approximation in Eq. (2.6) which was obtained by using straight quadratic interpolation.

For $\left.\frac{\partial^2 \Phi}{\partial x^2}\right|_P$, we consider the approximation in Eq. (4.26) to formulate

$$\begin{aligned}
\left.\frac{\partial^2 \Phi}{\partial x^2}\right|_P &= \frac{\partial}{\partial x}\left(\left.\frac{\partial \Phi}{\partial x}\right|_P\right) \approx \frac{1}{\Delta x}\left(\left.\frac{\partial \Phi}{\partial x}\right|_e - \left.\frac{\partial \Phi}{\partial x}\right|_w\right) \\
&= \frac{1}{\Delta x}\left[\beta_e \frac{\Phi_E - \Phi_P}{\delta_e} - \beta_w \frac{\Phi_P - \Phi_W}{\delta_w}\right] \\
&= \frac{2}{\delta_e + \delta_w}\left[\beta_e \frac{\Phi_E - \Phi_P}{\delta_e} - \beta_w \frac{\Phi_P - \Phi_W}{\delta_w}\right] \\
&= \frac{2}{\delta_e + \delta_w}\left[\frac{\beta_w}{\delta_w}\Phi_W - \left(\frac{\beta_e}{\delta_e} + \frac{\beta_w}{\delta_w}\right)\Phi_P + \frac{\beta_e}{\delta_e}\Phi_E\right] \\
&= \frac{2\beta_w}{\delta_w(\delta_e + \delta_w)}\Phi_W - \frac{2\beta_w(\beta_e\delta_w + \beta_w\delta_e)}{\delta_e\delta_w(\delta_e + \delta_w)}\Phi_P + \frac{2\beta_e}{\delta_e(\delta_e + \delta_w)}\Phi_E
\end{aligned} \tag{4.46}$$

If we set β_e and β_w to unity in Eq. 4.46, we obtain the approximation in Eq. (2.20), which was obtained by straight quadratic interpolation.

This provides an insight into the difference between a straight quadratic interpolation and the 'weighted' averaging interpolation. In both of the approximations we use three functional values. The difference is in the choice of the weighting factors.

The weighted averaging interpolation is named as the 'weighted approximation' in the literature (see Appendix A) and is almost classical for the solution of field problems.

Chapter 5
Preparations for Solution

Before commencing with the setting up of the block coupled system to be utilized later on, there are three basic steps to be completed.

Firstly, we note that the discretized equations constructed in the previous chapter are written only for one grid point, but are applicable to every point in the whole of the solution space. Therefore, it is necessary to apply these to each grid point interior to the solution region.

Secondly, it is well known that a physical fluid flow problem is not only characterized by the governing equations, but also complemented by certain boundary conditions to be applied along with these equations. The boundary conditions must be reflected into the equations for the points neighboring a boundary of the solution region. Boundary condition implementation, especially in the context of the coupled solution strategy introduced in Chap. 7, comprises a very important stage in the solution process. Therefore, it is necessary to examine the ways of implementing the boundary conditions of varying types efficiently and correctly into the discretized forms of the equations.

Finally, considering the nonlinear structure of the governing equations and the iterative nature of the solution procedures used, certain relaxation need to be implemented onto the discretized equations.

After these steps have been completed, all of the equations will then be assembled into a general matrix system to be solved by a suitable procedure.

The aim in this chapter is to introduce the ways of implementing certain preparatory processes in order to address these issues.

5.1 The Solution Region

The solution region of a certain physical problem is the area within which the governing equations are valid. For our purposes here, we will assume that such a region can be covered by a rectangular area, or can be regarded as a subset of such

© Springer International Publishing Switzerland 2016

Z. Mazhar, *Fully Implicit, Coupled Procedures in Computational Fluid Dynamics*,
Fluid Mechanics and Its Applications 115, DOI 10.1007/978-3-319-29895-5_5

Fig. 5.1 A rectangular area of solution

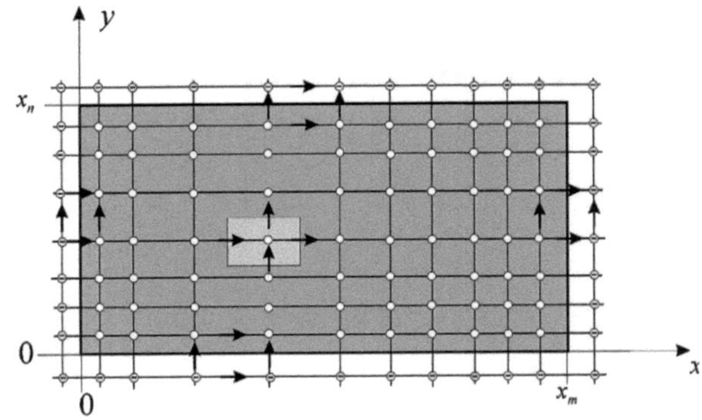

an area. Without loss of generality, we will assume that we are given the region defined by $R = \{(x, y\}, 0 \leq x \leq x_m, 0 \leq y \leq y_n)$ for some positive values x_m and y_n as shown in Fig. 5.1.

The solution region is subdivided into a finite number of rectangular subintervals, named as 'mass control volumes'. The grid system is formed in such a way that the 'real' boundary lines on the east, west, north and south boundary lines pass through midway from the two outermost lines of the grid nodes denoted by small circles. We suppose that the number of nodes, denoted by empty circles, in the x and y directions of the solution region are m and n, respectively. This means that there are a total of $N = m \times n$ inner nodes within the region altogether. The nodes outside the solution region marked with a minus sign are 'fictitious' nodes and they are not included in the set of actual nodes. Therefore, no equations are included for the variables at those nodes.

Now the set of discretized forms of the governing equations, Eqs. (4.32), (4.39) and (4.44) should be applied to all of these nodes using the control volumes defined in Figs. 4.3, 4.4 and 4.5, respectively. There is no need for writing any equation for the fictitious points. This will be handled while imposing the relevant boundary conditions on each boundary as will be shown in the following section. Since we will have N equations for each of the variables u, v and p, the total number of variables will be $3N$.

5.2 Boundary Conditions

Boundary conditions are necessary to describe certain physical conditions for the fluid flow problem under consideration. These conditions must be adequately interpreted and applied in the discretized forms of the governing differential equations.

The application of the boundary conditions for the dependent variables is a bit different from that used for equations which contain only one set of dependent variables of the same kind. Hence it is anticipated that some difficulties may be realized in applying the boundary conditions within the context of our coupled primitive system.

The aim in this section is to provide the readers with sufficient information on the correct application of the boundary conditions. In the subsections that follow, we first give a description of how a general boundary condition is applied. Following this, we discuss the implementation on various special cases.

Considering the importance of the correct implementation of the boundary conditions, the reader is urged to go through these sections thoroughly. Later on, the sections may serve as a reference when need arises.

5.2.1 *Types of Boundary Conditions*

Boundary conditions in a fluid flow are necessary in order to impose certain physical constraints describing, for example, obstacles around or within the flow region, enforced inflow or outflow velocity profiles or free surfaces. Some types of obstacles are 'building blocks', 'wind-breaks (shelterbelts)' and 'steps' as shown Fig. 5.2.

Such obstacles are defined by setting the velocity components u and v equal to zero at the surfaces of the obstacles.

Examples for a constant velocity profile are the inflow and outflow velocities $u = u_{in}$ and $v = u_{out}$, at the inlet and outlet openings of a 'Tank problem' shown in Fig. 5.3.

An enforced 'wind flow' profile, say for air with density of $\rho = 1.18\,\text{kg/m}^3$ and at 15 °C may be described by the formula $u_{in} = \frac{u_f}{\kappa}\ln\left(\frac{y}{y_0}\right)$, where $u_f = 0.31\,\text{m/sec}$ is the friction velocity, $\kappa = 0.4$ is the von Karman's constant, $y_0 = 0.007\,\text{m}$ is the roughness parameter and y is the height above the ground, is shown in Fig. 5.4.

A typical free surface is the east side or 'downstream' if the side is chosen far enough from the areas of any velocity changes. Such surfaces are characterized mathematically by imposing the conditions $\frac{\partial u}{\partial x} = 0$ and $\frac{\partial v}{\partial y} = 0$ at those surfaces. Furthermore, at the north side of such a region, conditions for the velocity components must be set to $u = u_{in}$ and $\frac{\partial v}{\partial y} = 0$.

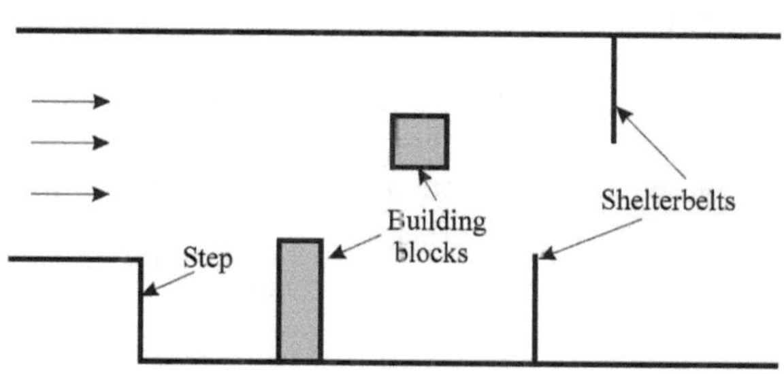

Fig. 5.2 Types of obstacles in a flow field

Fig. 5.3 Inlet and outlet boundary conditions

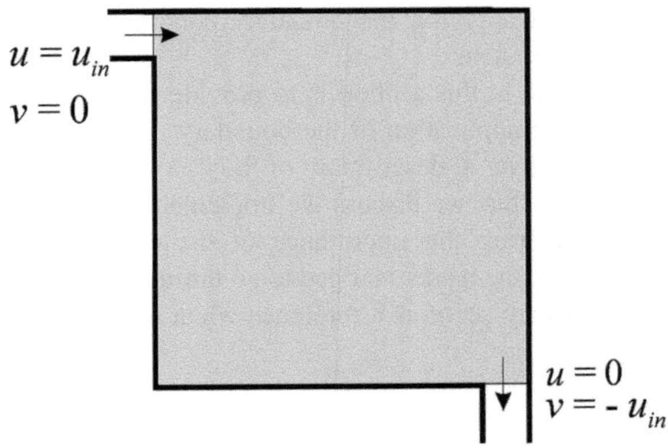

Fig. 5.4 Enforced wind flow and free surface flow

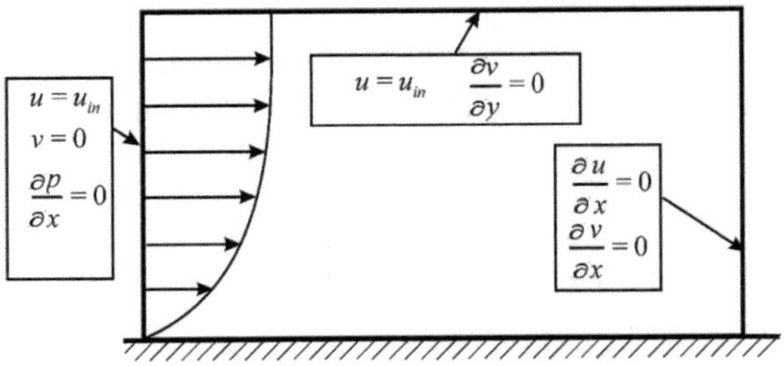

In general, a physical boundary condition can be described by the general formula

$$A\Phi = B\frac{\partial \Phi}{\partial n} + C, \tag{5.1}$$

in which *A, B* and *C* are constants, *n* may stand either for *x* or *y*, and Φ may stand for either *u*, *v* or *p*. If $B = 0$, the condition is said to be a 'Dirichlet' condition, if $A = 0$, the condition is called a 'Neumann' condition, otherwise it is called a 'mixed' type.

Suppose we have a general formula describing the behavior of Φ at a point *P*, neighboring the east face, shown with its neighboring points *S, W, E* and *N* in Fig. 5.5.

$$A_S\Phi_S + A_W\Phi_W + A_P\Phi_P + A_E\Phi_E + A_N\Phi_N = b_P \tag{5.2}$$

The boundary condition (5.1) for this case is written as

$$A\Phi = B\frac{\partial \Phi}{\partial x} + C \tag{5.3}$$

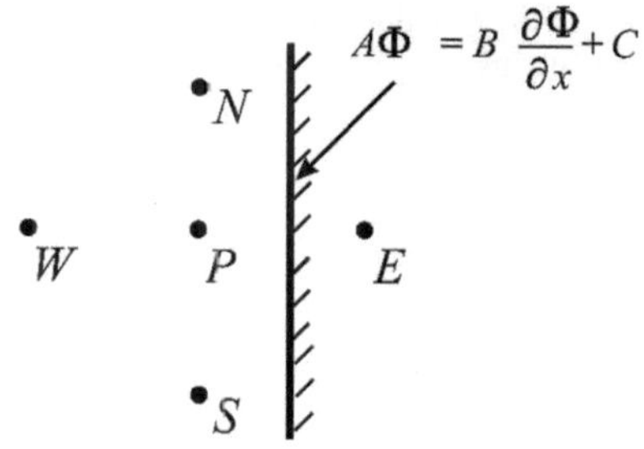

Fig. 5.5 General boundary conditions at a surface

First we write an approximation for (5.3) at the point P as

$$A\Phi_P = B\frac{\Phi_E - \Phi_P}{\Delta x} + C \tag{5.4}$$

Solving Φ_E in terms of Φ_P from (5.4) we get

$$\Phi_E = F + G\Phi_P \tag{5.5}$$

in which $F = -CD$ and $G = DA + 1$, where $D = \Delta x/B$.

Now substituting Φ_E from Eq. (5.5) into Eq. (5.2) gives

$$A_S\Phi_S + A_W\Phi_W + A_P\Phi_P + A_E(F + G\Phi_P) + A_N\Phi_N = b_P \tag{5.6}$$

or

$$A_S\Phi_S + A_W\Phi_W + (A_P + A_EG)\Phi_P + (0)\Phi_E + A_N\Phi_N = b_P - A_EF \tag{5.7}$$

The computational application of this change for the equation of Φ_P can now be done in three steps as

1. Update A_P by $A_P \leftarrow A_P + A_EG$
2. Update b_P by $b_P \leftarrow b_P - A_EF$
3. Set $A_E \leftarrow 0$.

5.2.2 Application of Various Types of Boundary Conditions into the Momentum Equations

Since the discretized forms of the two momentum equations are similar, let Φ stand either for u or v. Then an equation for Φ can be written as

$$A_S\Phi_S + A_W\Phi_W + A_P\Phi_P + A_E\Phi_E + A_N\Phi_N + A^p_{W(S)}p_P + A^p_{E(N)}p_E = b_P \tag{5.8}$$

Type 1: No slip type
Consider the boundary condition on a south side boundary as shown in Fig. 5.6.

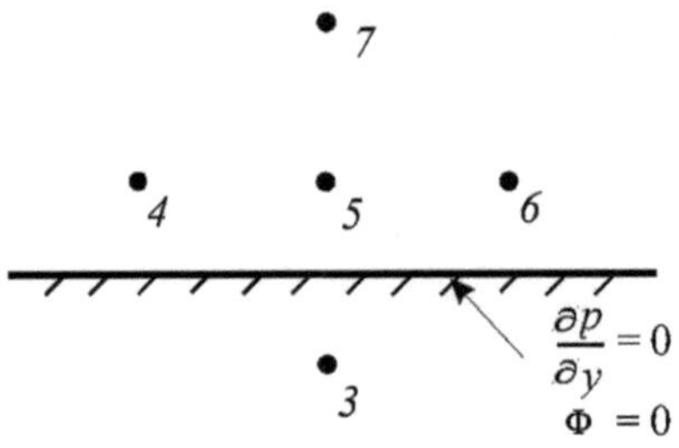

Fig. 5.6 No-slip boundary conditions

Considering the numbering of the nodes, Eq. (5.8) can be written, by using a simple change of notation, as

$$A_{53}\Phi_3 + A_{54}\Phi_4 + A_{55}\Phi_5 + A_{56}\Phi_6 + A_{57}\Phi_7 + A^p_{53}p_3 + A^p_{55}p_5 = b_5 \quad (5.9)$$

Here two conditions are demanded to be imposed. These are $\Phi_3 = -\Phi_5$ for Φ and $p_5 = p_3$ for p. Replacing Φ_3 by $-\Phi_5$ and p_3 by p_5 in Eq. (5.9) gives

$$A_{53}(-\Phi_5) + A_{54}\Phi_4 + A_{55}\Phi_5 + A_{56}\Phi_6 + A_{57}\Phi_7 + A^p_{53}(p_5) + A^p_{55}p_5 = b_5 \quad (5.9')$$

To impose the condition on Φ, A_{55} is first replaced by $A_{55} - A_{53}$, followed by setting A_{53} to zero. The condition on p is imposed by first replacing A^p_{55} by $A^p_{55} + A^p_{53}$, followed by setting A^p_{53} to zero. Equation (5.9) then becomes

$$(0)\Phi_3 + A_{54}\Phi_4 + (A_{55} - A_{53})\Phi_5 + A_{56}\Phi_6 + A_{57}\Phi_7 + (0)p_3 + (A^p_{55} + A^p_{53})p_5 = b_5 \quad (5.10)$$

The implementation of this process can be summarized as follows:

1. Update A_{55} by $A_{55} \leftarrow A_{55} - A_{53}$
2. Update A^p_{55} by $A^p_{55} \leftarrow A^p_{55} - A^p_{53}$
3. Set $A_{53} \leftarrow 0$
4. Set $A^p_{53} \leftarrow 0$

Type 2: Gradient type boundary condition
This type of a condition is shown in Fig. 5.7 at a north side boundary.

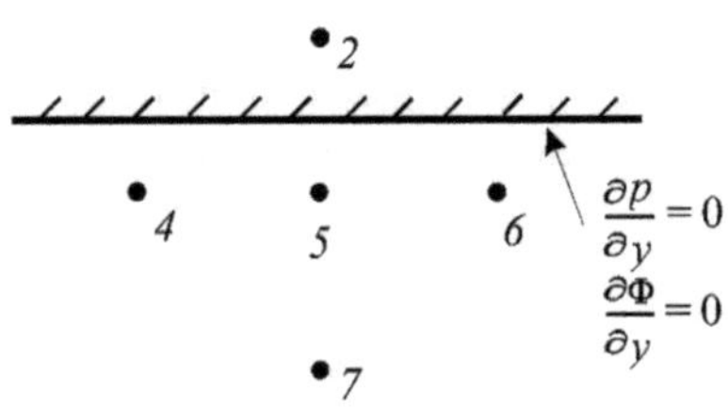

Fig. 5.7 Gradient type boundary conditions

Considering again the numbering of the nodes, Eq. (5.8) can be written as

$$A_{57}\Phi_7 + A_{54}\Phi_4 + A_{55}\Phi_5 + A_{56}\Phi_6 + A_{52}\Phi_2 + A^p_{55}p_5 + A^p_{52}p_2 = b_5 \quad (5.11)$$

Here again two conditions are demanded to be imposed. These are $\Phi_2 = \Phi_5$ for Φ and $p_5 = p_2$ for p. To impose the condition on Φ, A_{55} is first replaced by $A_{55} + A_{52}$, followed by setting A_{52} to zero. The condition for p is imposed by first replacing A^p_{55} by $A^p_{55} + A^p_{52}$, followed by setting A^p_{52} to zero. Equation (5.11) then becomes

$$(0)\Phi_7 + A_{54}\Phi_4 + (A_{55} + A_{57})\Phi_5 + A_{56}\Phi_6 + A_{52}\Phi_5 + (A^p_{55} + A^p_{52})p_5 + (0)p_2 = b_5 \quad (5.12)$$

5.2.3 *Application of Boundary Conditions into the Coupled Equations*

Now we turn our attention to the discretized momentum and mass conservation equations in (4.32), (4.39) and (4.44). A boundary condition set for a point neighboring a boundary must be imposed on the whole set of these equations.

If we had an explicit equation for pressure, we would also need to impose certain conditions on these equations as well. Considering the relativeness of the pressure values in an incompressible flow problem, the only condition to be imposed is to set the pressure value at only one reference point in the solution domain, while letting the pressures be 'free' by specifying $\frac{\partial p}{\partial n} = 0$ at the boundaries. Surprisingly enough, in contrast to certain procedures described in the literature in which special 'equations' are derived for pressure, this last condition need not even be set in the special solution procedures described in this book, since there is no explicit equation for pressure. Pressure boundary conditions are only reflected in the pressure terms of the two momentum equations as discussed below.

Boundary conditions for the velocity components must be imposed not only on the momentum equations, but also on the mass conservation equations.

We will consider various conditions, but it must hereby be stated that a certain combination of these conditions may appear in actual flow problems. Nevertheless, the examples presented give the reader all the knowledge on how other combinations will be treated.

Case 1: South face Consider the boundary conditions on the south side of the rectangular region as shown in Fig. 5.8.

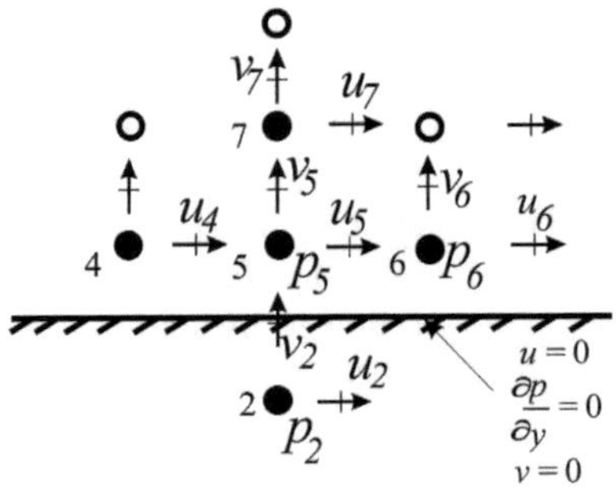

Fig. 5.8 South-face boundary conditions

The discretized mass and momentum equations for the point denoted by (5) can be written as follows.

$$
\begin{aligned}
&A^u_{52}u_2 + A^u_{54}u_4 + A^u_{55}u_5 + A^u_{56}u_6 + A^u_{57}u_{57} + A^{pu}_{55}p_5 + A^{pu}_{56}p_6 = b^u_5 \\
&A^v_{52}v_2 + A^v_{54}v_4 + A^v_{55}v_5 + A^v_{56}v_6 + A^v_{57}v_7 + A^{pv}_{55}p_5 + A^{pv}_{57}p_7 = b^v_5 \\
&A^{Cu}_4 u_4 + A^{Cu}_{55}u_5 + A^{Cv}_{52}v_2 + A^{Cv}_{55}v_5 = 0
\end{aligned}
$$

Conditions $\frac{\partial p}{\partial y} = 0$, $u = 0$ and $v = 0$ on the boundary can be expressed as $p_5 = p_2$, $u_5 = -u_2$ and $v_2 = 0$, respectively. Here nothing has to be done for the pressure condition since p_2 does not appear in the equations. Only the conditions for the velocities are applied onto the momentum equations as well as onto the mass conservation equation by updating the above equations as follows:

$$
\begin{aligned}
&(0)u_2 + A^u_{54}u_4 + (A^u_{55} - A^u_{52})u_5 + A^u_{56}u_6 + A^u_{57}u_{57} + A^{pu}_{55}p_5 + A^{pu}_{56}p_6 = b^u_5 \\
&(0)v_2 + A^v_{54}v_4 + A^v_{55}v_5 + A^v_{56}v_6 + A^v_{57}v_7 + A^{pv}_{55}p_5 + A^{pv}_{57}p_7 = b^v_5 \\
&A^{Cu}_4 u_4 + A^{Cu}_{55}u_5 + (0)v_2 + A^{Cv}_{55}v_5 = 0
\end{aligned}
$$

The coding for these changes is implemented as follows:

$$
\begin{aligned}
&A^u_{55} \leftarrow A^u_{55} - A^u_{52} \\
&A^u_{52} \leftarrow 0 \\
&A^v_{55} \leftarrow A^v_{55} - A^v_{52} \\
&A^v_{52} \leftarrow 0 \\
&A^{cv}_{52} \leftarrow 0
\end{aligned}
$$

Case 2: East face Consider the conditions given on the east face as shown in Fig. 5.9. The discretized mass and momentum equations for the point denoted by (3) can be written as follows.

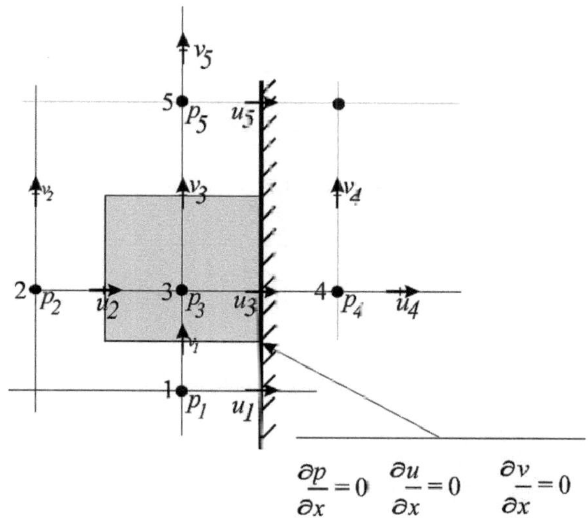

Fig. 5.9 East-face boundary conditions

$$
\begin{aligned}
&A^{u}_{31}u_1 + A^{u}_{32}u_2 + A^{u}_{33}u_3 + A^{u}_{34}u_4 + A^{u}_{35}u_5 + A^{pu}_{33}p_3 + A^{pu}_{34}p_4 = b^{u}_3\\
&A^{v}_{31}v_1 + A^{v}_{32}v_2 + A^{v}_{33}v_3 + A^{v}_{34}v_4 + A^{v}_{35}v_5 + A^{pv}_{33}p_3 + A^{pv}_{35}p_5 = b^{v}_3\\
&A^{Cu}_{32}u_2 + A^{Cu}_{33}u_3 + A^{Cv}_{31}v_1 + A^{Cv}_{33}v_3 = 0
\end{aligned}
$$

Conditions $\frac{\partial p}{\partial x} = 0$, $\frac{\partial u}{\partial x} = 0$ and $\frac{\partial v}{\partial x} = 0$ on the boundary can be expressed as $p_4 = p_3$, $u_4 = u_3$ and $v_4 = v_3$, respectively. These conditions are applied by updating the above equations as follows:

$$
\begin{aligned}
&A^{u}_{31}u_1 + A^{u}_{32}u_2 + (0)u_3 + (A^{u}_{34} + A^{u}_{33})u_4 + A^{u}_{35}u_5 + (A^{pu}_{33} + A^{pu}_{34})p_3 + (0)p_4 = b^{u}_3\\
&A^{v}_{31}v_1 + A^{v}_{32}v_2 + (0)v_3 + (A^{v}_{34} + A^{v}_{33})v_4 + A^{v}_{35}v_5 + A^{pv}_{33}p_3 + A^{pv}_{35}p_5 = b^{v}_3\\
&A^{Cu}_{32}u_2 + A^{Cu}_{33}u_3 + A^{Cv}_{31}v_1 + A^{Cv}_{33}v_3 = 0
\end{aligned}
$$

Case 3: North face Consider the conditions given on the north face as shown in Fig. 5.10. The discretized mass and momentum equations for the point denoted by (5) can be written as follows.

$$
\begin{aligned}
&A^{u}_{52}u_2 + A^{u}_{54}u_4 + A^{u}_{55}u_5 + A^{u}_{56}u_6 + A^{u}_{57}u_7 + A^{pu}_{55}p_5 + A^{pu}_{56}p_6 = b^{u}_5\\
&A^{v}_{52}v_5 + A^{v}_{54}v_4 + A^{v}_{55}v_5 + A^{v}_{56}v_6 + A^{v}_{57}v_7 + A^{pv}_{55}p_5 + A^{pv}_{57}p_7 = b^{v}_5\\
&A^{Cu}_{54}u_4 + A^{Cu}_{55}u_5 + A^{Cv}_{52}v_2 + A^{Cv}_{55}v_5 = 0
\end{aligned}
$$

Conditions $\frac{\partial p}{\partial y} = 0$, and $\frac{\partial v}{\partial y} = 0$ on the boundary can be expressed as $p_5 = p_7$ and $v_7 = v_5$, respectively. The condition $u = u_{wall}$ is different. Considering that the points (7) and (5) are equidistant from the wall face, the relationship $u_{wall} = (u_7 + u_5)/2$, or $u_7 = 2u_{wall} - u_5$ can be used. Applying these conditions to the above equations give

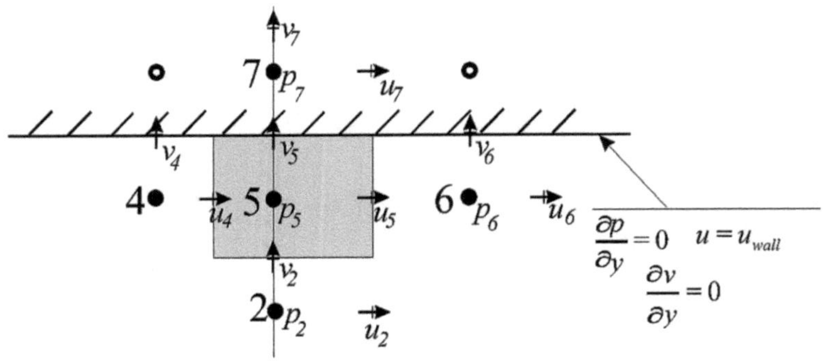

Fig. 5.10 North-face boundary conditions

$$A^u_{52}u_2 + A^u_{54}u_4 + (A^u_{55} - A^u_{57})u_5 + A^u_{56}u_6 + (0)u_7 + A^{pu}_{55}p_{55} + A^{pu}_{56}p_6 = b^u_5 - 2A^u_{57}u_{wall}$$
$$A^v_{52}v_5 + A^v_{54}v_4 + (A^v_{55} + A^v_{57})v_5 + A^v_{56}v_6 + (0)v_7 + (A^{pv}_{55} + A^{pv}_{57})p_5 + (0)p_7 = b^v_5$$
$$A^{Cu}_{54}u_4 + A^{Cu}_{55}u_5 + A^{Cv}_{52}v_2 + A^{Cv}_{55}v_5 = 0$$

The algorithm for the implementation is obvious to the reader by now.

Case 4: West face Consider the conditions given on the west face as shown in Fig. 5.11. The discretized mass and momentum equations for the point denoted by (3) can be written as follows.

Fig. 5.11 West-face boundary conditions

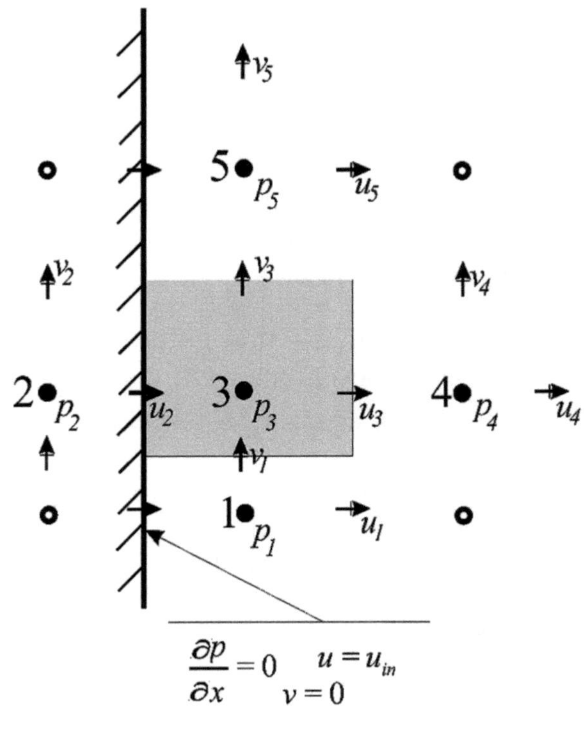

$$A^u_{31}u_1 + A^u_{32}u_2 + A^u_{33}u_3 + A^u_{34}u_4 + A^u_{35}u_5 + A^{pu}_{33}p_3 + A^{pu}_{34}p_4 = b^u_3$$
$$A^v_{31}v_1 + A^v_{32}v_2 + A^v_{33}v_3 + A^v_{34}v_4 + A^v_{35}v_3 + A^{pv}_{33}p_3 + A^{pv}_{35}p_5 = b^v_3$$
$$A^{Cu}_{32}u_2 + A^{Cu}_{33}u_3 + A^{Cv}_{31}v_1 + A^{Cv}_{33}v_3 = 0$$

Conditions $\frac{\partial p}{\partial x} = 0$, $u = u_{in}$ and $v = 0$ on the boundary can be expressed as $p_2 = p_3$, $u_2 = u_{in}$ and $v_2 = -v_3$, respectively. To reflect these conditions, the above equations should now be updated to as

$$A^u_{31}u_1 + (0)u_2 + A^u_{33}u_3 + A^u_{34}u_4 + A^u_{35}u_5 + A^{pu}_{33}p_3 + A^{pu}_{34}p_4 = b^u_3 - A^u_{32}u_{in}$$
$$A^v_{31}v_1 + (0)v_2 + (A^v_{33} - A^v_{32})v_3 + A^v_{34}v_4 + A^v_{35}v_3 + A^{pv}_{33}p_3 + A^{pv}_{35}p_5 = b^v_3$$
$$(0)u_2 + A^{Cu}_{33}u_3 + A^{Cv}_{31}v_1 + A^{Cv}_{33}v_3 = -A^{Cu}_{32}u_{in}$$

The coding of the implementation is as follows:

$$b^u_3 \leftarrow b^u_3 - A^u_{32}u_{in}$$
$$A^u_{32} \leftarrow 0$$
$$A^v_{33} \leftarrow A^v_{33} - A^v_{32}$$
$$A^v_{32} \leftarrow 0$$
$$b^p_3 \leftarrow b^p_3 - A^{Cu}_{32}u_{in}$$
$$A^{Cu}_{32} \leftarrow 0$$

Case 5: Boundary conditions at an inlet Consider a typical inlet corner and suitable conditions as shown in Fig. 5.12.

The discretized mass and momentum equations for the point denoted by (3) can be written as follows.

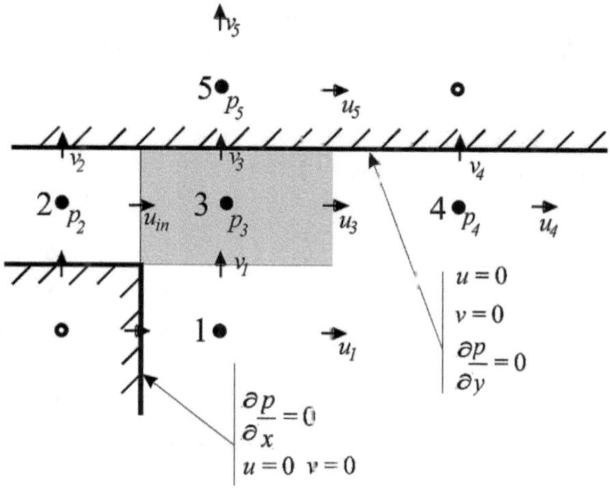

Fig. 5.12 Boundary conditions at an inlet

$$A^{u}_{31}u_1 + A^{u}_{32}u_2 + A^{u}_{33}u_3 + A^{u}_{34}u_4 + A^{u}_{35}u_5 + A^{pu}_{33}p_3 + A^{pu}_{34}p_4 = b^{u}_{3}$$
$$A^{v}_{31}v_1 + A^{v}_{32}v_2 + A^{v}_{33}v_3 + A^{v}_{34}v_4 + A^{v}_{35}v_3 + A^{pv}_{33}p_3 + A^{pv}_{35}p_5 = b^{v}_{3}$$
$$A^{Cu}_{32}u_2 + A^{Cu}_{33}u_3 + A^{Cv}_{31}v_1 + A^{Cv}_{33}v_3 = 0$$

The conditions given can be represented as $u_3 = -u_5$, $v_3 = 0$, $p_5 = p_3$, $p_2 = p_3$, and $u_2 = u_{in}$. The above equations should now be updated to reflect these conditions as

$$A^{u}_{31}u_1 + (0)u_2 + (A^{u}_{33} - A^{u}_{35})u_3 + A^{u}_{34}u_4 + (0)u_5 + A^{pu}_{33}p_3 + A^{pu}_{34}p_4 = b^{u}_{3} - A^{u}_{32}u_{in}$$
$$(0)v_1 + (0)v_2 + (1)v_3 + (0)v_4 + (0)v_3 + (0)p_3 + (0)p_5 = 0$$
$$(0)u_2 + A^{Cu}_{33}u_3 + A^{Cv}_{31}v_1 + (0)v_3 = -A^{Cu}_{32}u_{in}$$

It is noteworthy here that with the application of the boundary conditions of Type 4 and 5 in the discretized form of the mass conservation equation, the right hand side of the equations becomes nonzero. Therefore we must rewrite the general form of Eq. (4.44) for mass conservation at a grid point P as

$$A^{Cu}_{W}u_W + A^{Cu}_{P}u_P + A^{Cv}_{S}v_S + A^{Cv}_{P}v_P = b^{p}_{P} \tag{5.13}$$

and apply the updates as usual.

Application of such a condition at an outlet can be accomplished in a similar manner. There may be, of course, more number of grid points at an inlet or at an outlet. In such a case, the process must be repeated for each such point.

Case 6: Boundary conditions at a wind-break We consider the no-slip type boundary conditions on a wind-break as shown in Fig. 5.13.

Application of these boundary conditions is much different from the two cases discussed above. In those cases, the 'outer' nodes were not actually used, that is to say, their equations were not constructed in the overall formulation of the difference equations. In this case the difference equations are formed for both of the involved nodes numbered 4 and 5. The variable on the right hand side of node (4) is

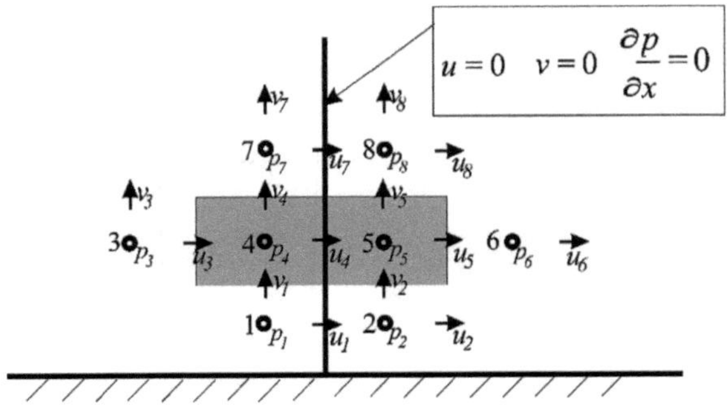

Fig. 5.13 Boundary conditions at a wind-break

'fictitious' for the variable at the node (5), and the variable on the left hand side of node (5) is 'fictitious' for the variable at the node (4). Therefore, the variables at the nodes (4) and (5) must be completely isolated from the effects of each other by updating their corresponding difference equations. Denoting the 'fictitious' nodes by (4′) and (5′) in a respective form, the difference equations at the nodes (4) and (5) are written as

For node (4)

$$A^u_{41}u_1 + A^u_{43}u_3 + A^u_{44}u_4 + A^u_{45'}u_{5'} + A^u_{47}u_7 + A^{pu}_{44}p_4 + A^{pu}_{45'}p_{5'} = b^u_4$$
$$A^v_{41}v_1 + A^v_{43}v_3 + A^v_{44}v_4 + A^v_{45'}v_{5'} + A^v_{47}v_7 + A^{pv}_{44}p_4 + A^{pv}_{45'}p_{5'} = b^v_4$$
$$A^{Cu}_{43}u_3 + A^{Cu}_{44}u_4 + A^{Cv}_{41}v_1 + A^{Cv}_{47}v_7 = 0$$

For node (5)

$$A^u_{52}u_5 + A^u_{54'}u_{4'} + A^u_{55}u_5 + A^u_{56}u_6 + A^u_{58}u_8 + A^{pu}_{55}p_5 + A^{pu}_{56}p_6 = b^u_5$$
$$A^v_{52}v_2 + A^v_{54'}v_{4'} + A^v_{55}v_5 + A^v_{56}v_6 + A^v_{58}v_8 + A^{pv}_{55}p_5 + A^{pv}_{58}p_8 = b^v_5$$
$$A^{Cu}_{54'}u_{4'} + A^{Cu}_{55}u_5 + A^{Cv}_{52}v_2 + A^{Cv}_{55}v_5 = 0$$

To impose the conditions given as $u = 0$, $v = 0$ and $\frac{\partial p}{\partial x} = 0$ in Fig. 5.13, we must impose $u_4 = 0$, $v_4 = -v_5$ and $p_4 = p_5$ onto the above two sets of equations as follows:

For node (4)

$$(0)u_1 + (0)u_3 + (1)u_4 + (0)u_{5'} + (0)u_7 + (0)p_4 + (0)p_{5'} = (0)$$
$$A^v_{41}v_1 + A^v_{43}v_3 + (A^v_{44} - A^v_{45'})v_4 + (0)v_{5'} + A^v_{47}v_7 + (A^{pv}_{44} + A^{pv}_{45'})p_4 + (0)p_{5'} = b^v_4$$
$$A^{Cu}_{43}u_3 + (0)u_4 + A^{Cv}_{41}v_1 + A^{Cv}_{47}v_7 = 0$$

For node (5)

$$A^u_{52}u_5 + (0)u_{4'} + A^u_{55}u_5 + A^u_{56}u_6 + A^u_{58}u_8 + A^{pu}_{55}p_5 + A^{pu}_{56}p_6 = b^u_5$$
$$A^v_{52}v_2 + (0)v_{4'} + (A^v_{55} - A^v_{54'})v_5 + A^v_{56}v_6 + A^v_{58}v_8 + A^{pv}_{55}p_5 + A^{pv}_{58}p_8 = b^v_5$$
$$(0)u_{4'} + A^{Cu}_{55}u_5 + A^{Cv}_{52}v_2 + A^{Cv}_{55}v_5 = 0$$

This completes the process.

Case 7: Treatment for the nodes within a building block There may be cases in which a 'building block' is placed in the flow field under consideration, as shown in Fig. 5.14.

In such a case, the discretizated set of equations for the variables at the nodes adjacent to the boundary of the building block, but outside the block, are treated in a similar way as described in the above cases. The nodes which reside within the

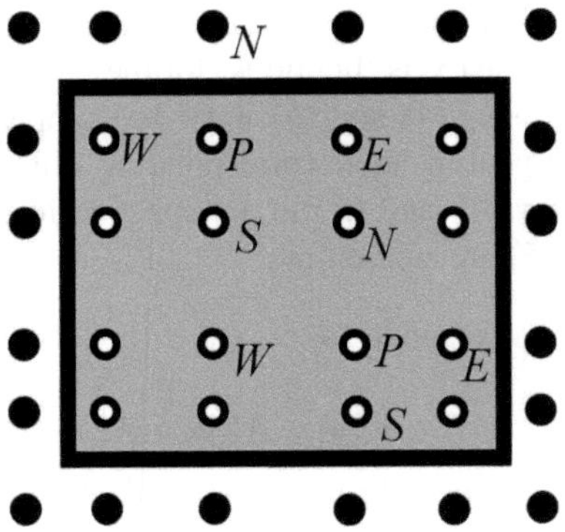

Fig. 5.14 Boundary conditions on a building block

block (empty circles) are actually outside of the solution region. These nodes are 'fictitious' to the respective inner nodes. However, the equations for these nodes still need to be kept in the set of overall discretized equations in order to be able to keep the integrity of the structured form of the resulting matrices to be constructed.

The discretized momentum equations for such a point, denoted by P, may be set as

$$(0)u_S + (0)u_W + (1)u_P + (0)u_E + (0)u_N + (0)p_P + (0)p_E = (0)$$
$$(0)v_S + (0)v_W + (1)v_P + (0)v_E + (0)v_N + (0)p_P + (0)p_N = (0)$$

There is no need to perform any changes to the mass conservation equation for the point P.

These equations are really 'dummy' equations. They only act to define the building block in the flow field. If a point-by-point implementation is used with a block coupled procedure, which is possible as we will see in Chap. 7, there will be no need for setting up of a 'dummy' equation for such points.

5.3 Incorporating Relaxation

The governing momentum equations pose very strong nonlinearities due to the product terms appearing in the first order derivatives of the velocity components. In some physical considerations, even these nonlinearities may appear in the derivative terms of the second order as well. These nonlinearities tend to undermine an iterative solution of the whole discretized system, especially in the very early stages of the iterations in which the changes in the dependent variables are high. To avoid divergence, these changes must be slowed down. This is accomplished by a process called *under-relaxation* which must be applied in the discretized forms of the momentum equations.

Under-relaxation should not be applied to the mass conservation equation.

Although there are various ways of applying under-relaxation, one convenient way will be described here and applied throughout the book. Let one of the discretized forms of the two momentum equations be represented in the compact form

$$A_P\Phi_P + \sum A_{nb}\Phi_{nb} = b \tag{5.14}$$

We write this equation in an alternate form as

$$\Phi_P = \frac{b - \sum A_{nb}\Phi_{nb}}{A_P} \tag{5.15}$$

Let Φ_P^* represent the value of Φ_P obtained from a previous iteration. Adding and subtracting Φ_P^* on the right hand side of Eq. (5.15) gives

$$\Phi_P = \Phi_P^* + \left(\frac{b - \sum A_{nb}\Phi_{nb}}{A_P} - \Phi_P^*\right) \tag{5.16}$$

The expression in the parenthesis represents the *change* in Φ_P which has been obtained in the previous iteration. The value of Φ_P can now be modified by multiplying the term in paranthesis by a constant factor α, called the *relaxation factor.* Equation (5.16) then becomes

$$\Phi_P = \Phi_P^* + \alpha\left(\frac{b - \sum A_{nb}\Phi_{nb}}{A_P} - \Phi_P^*\right) \tag{5.17}$$

which can be written as

$$\frac{A_P}{\alpha}\Phi_P = b - \sum A_{nb}\Phi_{nb} + (1 - \alpha)\frac{A_P}{\alpha}\Phi_P^* \tag{5.18}$$

This operation amounts to letting only a percentage of the changes to be reflected to the variable Φ_P. If $\alpha = 0.6$ for example, it will amount to letting only sixty percent of the changes to be reflected. This means a forty percent clip-off from the changes.

Now we write Eq. (5.18) in a more convenient form by letting $\alpha = \frac{E}{1+E}$, as follows

$$\frac{1+E}{E}A_P\Phi_P = b - \sum A_{nb}\Phi_{nb} + \frac{1}{E}A_P\Phi_P^* \tag{5.19}$$

This formulation of the relaxation process is called *the E-factor formulation* in the literature (see Appendix A).

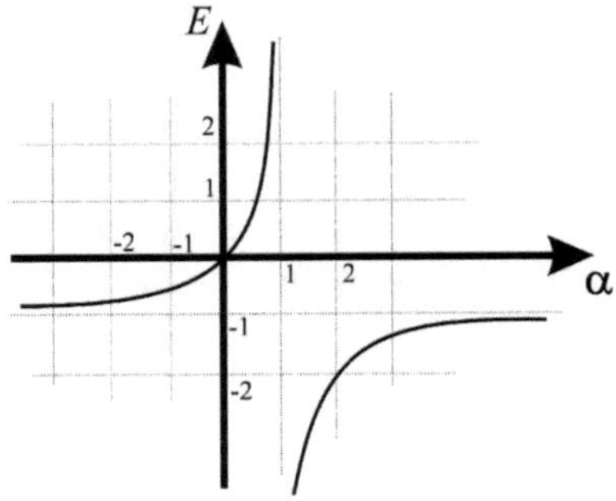

Fig. 5.15 Relaxation factors α and E

For the application of the above described relaxation, it suffices to update the central coefficient A_P by $\frac{1+E}{E}A_P$ and the corresponding right hand side coefficient b by $b - \frac{1}{E}A_P\Phi_P^*$ in each of the discretized forms of the momentum equations (Eqs. 4.32 and 4.39).

Noting that $E = \frac{\alpha}{1-\alpha}$, Fig. 5.15 represents a graphical relationship between α and E.

If $0 < \alpha \leq 1$ the relaxation is called under-relaxation. Under-relaxation helps in slowing down the rapid changes that may occur during an iterative process and avoids divergence. It can be seen from Fig. 5.15 that for under-relaxation, the E factor must be greater than zero. A very high value of E means that no relaxation is applied.

If $\alpha > 1$, the process is called *over-relaxation* and it helps in speeding up an already convergent iterative process. From Fig. 5.15 it is clear that over-relaxation cannot be applied with the E-factor formulation. Nevertheless, for incompressible fluid flow problems, only under-relaxation is necessary.

Chapter 6
Assembling the Discretized Equations into a Block Matrix System

In the previous chapter, a set of three equations for one grid point of the solution region was obtained. Equations of the same type apply as well for all the grid points in the region. Therefore the next step in the solution process is to use the general forms of the discretized equations obtained and write explicit linear algebraic equations for the unknown variables at all the grid points in the solution region. The step is completed by assembling all of these equations into a global block matrix form. This is accomplished in this chapter.

6.1 The Numbering Scheme

The numbering scheme plays some vital roles in the solution process. Firstly, it gives a knowledge of the location, and therefore, of the relationship between the neighboring points in the solution region. Secondly, it helps to correctly place the corresponding coefficients of the discretized equations into the respective places of the vectors forming the matrix system. Thirdly, and most importantly, it helps to shape up (or structure) the resulting matrices. This helps in determining a certain, efficient way for the solution of the resulting system. The structure of the resulting block system is very important in the solution process. Solution of linear unstructured systems is much more difficult and time consuming compared to the structured ones.

The boundary lines of the solution region are placed midway between the two outermost rectangular grid lines. Equations are written only for the points that lie inside of the rectangular boundary. The grid points which lie at the outside of the solution region are treated as 'fictitious' grid points. Therefore, only the 'inner' nodes in the solution region are to be numbered.

To obtain a useful structure for the resulting matrix, a suitable numbering strategy must be chosen. For our purposes here, among various types of strategies, we choose to begin numbering the nodes starting from the lower left corner and

© Springer International Publishing Switzerland 2016

Z. Mazhar, *Fully Implicit, Coupled Procedures in Computational Fluid Dynamics*,
Fluid Mechanics and Its Applications 115, DOI 10.1007/978-3-319-29895-5_6

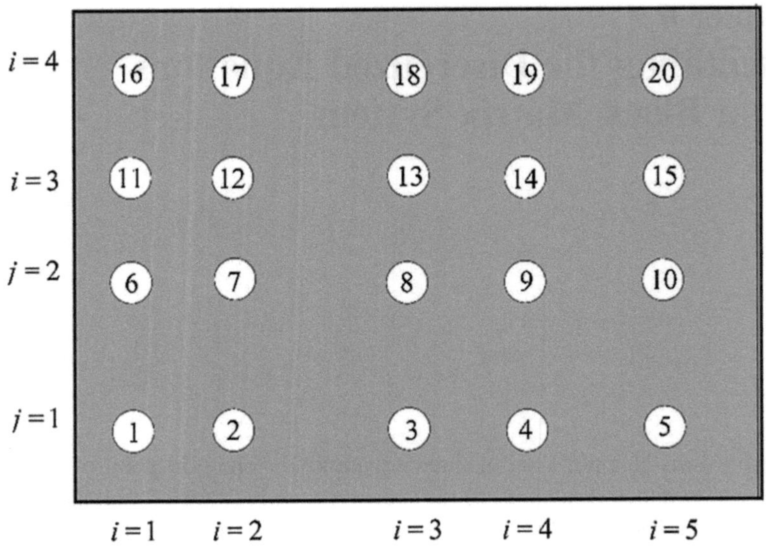

Fig. 6.1 Grid numbering strategy

continuing until the lower right corner. This completes the numbering of the nodes in the lower line. We continue this numbering, going line by line in the upper direction until completing numbering of the uppermost line.

Now let the number of nodes in one horizontal line be m, and the number of nodes in a vertical line be n. Therefore, there results a total number of $N = m \times n$ grid points. Let us number the horizontal and vertical lines by $j = 1, n$ and $i = 1, m$, respectively. Therefore the number of a nodal point in the solution region represented by the indices (i,j) is given by $k = N(i,j) = i + (j - 1) \times m$.

We demonstrate this strategy on an $(m, n) = (5, 4)$ grid as shown in Fig. 6.1. Here we have a total of $N = 20$ nodes. The four neighboring points at the south, west, east and north sides of a point numbered by k are numbered as $k - m$, $k - 1$, $k + 1$ and $k + m$, respectively. Furthermore, the point at the southeast of the point k is numbered by $k - m + 1$ and the one at the northwest is numbered by $k + m - 1$.

6.2 Construction of the Block Matrix System

We first write the u momentum equations starting from the point numbered by 1 and going up to N. We continue writing corresponding equations for the v momentum equations and then followed by the mass conservation equations, in the same manner. Now with a simple change of notation for the coefficients which will aid us in storing all the coefficients in vector form, the three equations for a node numbered, say k, will be written as

$$A^u_{S_k} u_{k-m} + A^u_{W_k} u_{k-1} + A^u_{P_k} u_k + A^u_{E_k} u_{k+1} + A^u_{N_k} u_{k+m} + A^p_{W_k} p_k + A^p_{E_k} p_{k+1} = b^u_k, \quad k = 1, N, \tag{6.1a}$$

$$A^v_{S_k} v_{k-m} + A^v_{W_k} v_{k-1} + A^v_{P_k} v_k + A^v_{E_k} v_{k+1} + A^v_{N_k} v_{k+m} + A^p_{S_k} p_k + A^p_{N_k} p_{k+m} = b^v_k, \quad k = 1, N, \tag{6.1b}$$

$$A^C_{W_k} u_{k-1} + A^C_{E_k} u_k + A^C_{S_k} v_{k-m} + A^C_{N_k} v_k = b^p_k, \quad k = 1, N, \tag{6.1c}$$

Now these equations are transferred into the appropriate places of a block matrix system whose structure is shown in Fig. 6.2.

The transfer starts from node 1 and goes up to node N.

The matrix that has been obtained in this way will be named to have the *primitive form*.

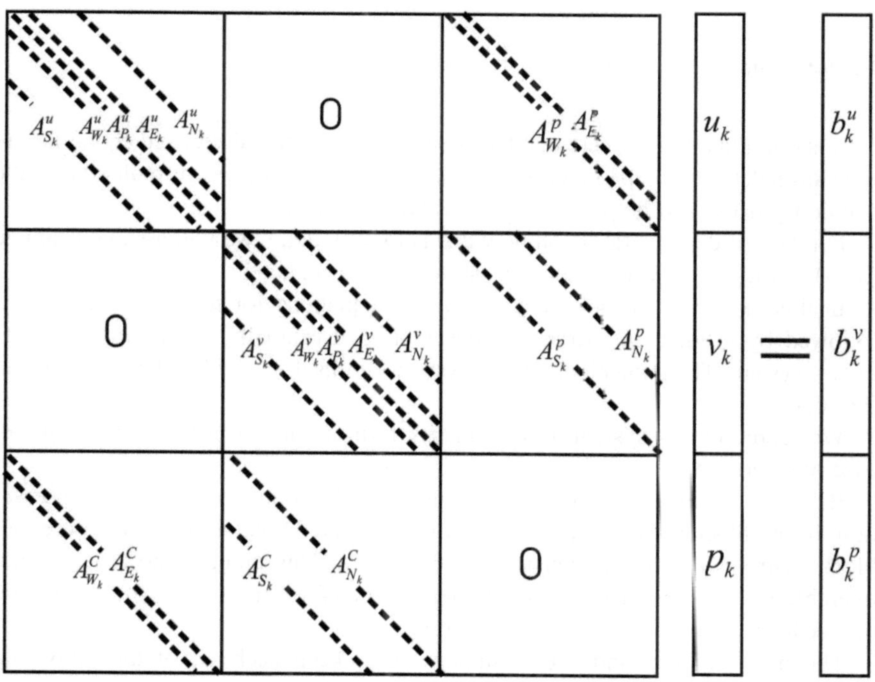

Fig. 6.2 The general block matrix system

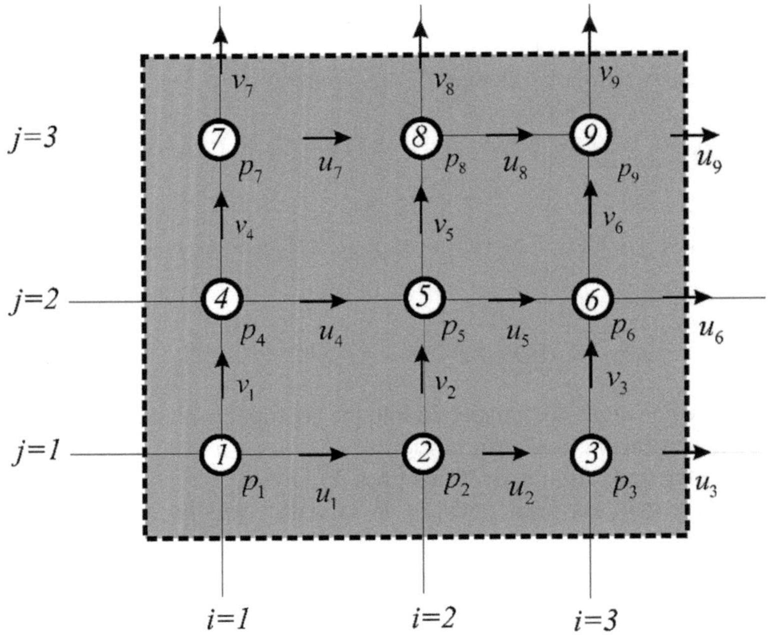

Fig. 6.3 A simple 3 × 3 grid

It is important to note here that the coefficients from the discretized equations are transformed to and kept in vector form. This will aid in the formulation of the solution process and help minimize the storage needed.

For ease of demonstration on how this is exactly done, we will assume a simple $(m, n) = (3, 3)$ grid with 9 nodes as shown in Fig. 6.3.

In this example, we note that we have 9 equations for u, 9 equations for v, followed by 9 equations for mass conservation. Eventually, we have a 27 × 27 matrix system. Then, the block matrix system for this 3 × 3 grid will be as shown in Fig. 6.4.

We represent this system in a submatrix form as shown in Fig. 6.5 for further ease of reference.

Before continuing any further, we must note the following important observation. In our block system, we have $3N$ equations and $3N$ unknowns. Therefore, our block matrix will have dimensions of $3N \times 3N$. This amount, however, is not tolerable even with small scaled problems. Therefore, a storage strategy in full matrix form is completely out of question.

The most efficient and useful strategy is to keep each set of the individual coefficients in vector form as shown in Fig. 6.4. The procedures that are presented in the next chapter are most suited for this vector representation.

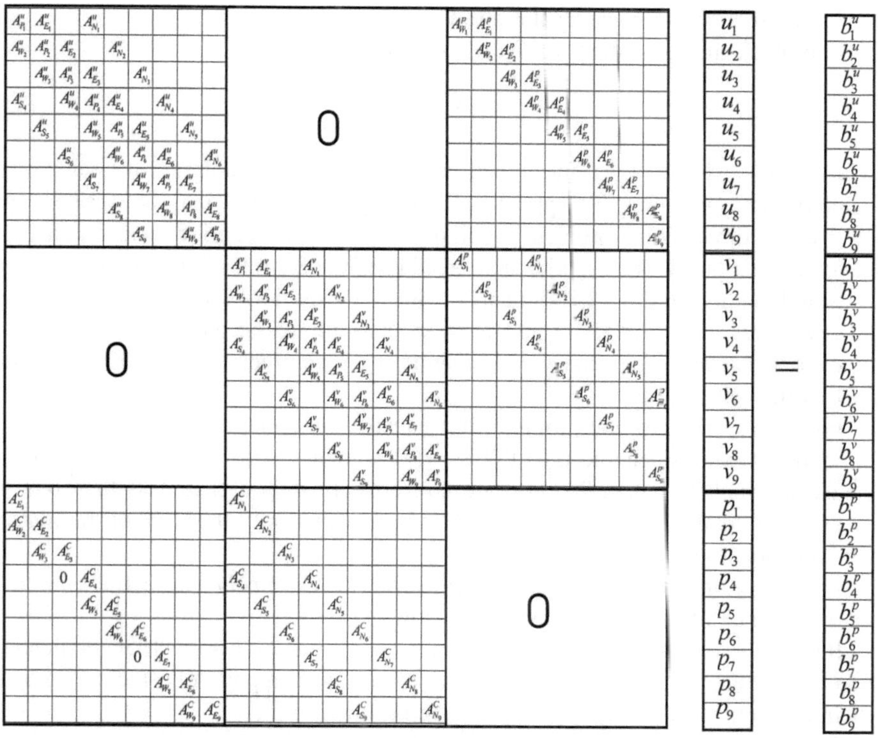

Fig. 6.4 Block 27 × 27 system for the 3 × 3 grid

Fig. 6.5 Submatrix form of the block matrix system

$$\begin{bmatrix} A_{11} & 0 & A_{13} \\ 0 & A_{22} & A_{23} \\ A_{31} & A_{32} & 0 \end{bmatrix} \begin{bmatrix} u \\ v \\ p \end{bmatrix} = \begin{bmatrix} b^u \\ b^v \\ b^p \end{bmatrix}$$

6.3 Disadvantages of the Block Matrix

In order to be able to devise an efficient solution procedure for the linear system of equations obtained, we need to know the properties of the coefficient matrix. In this regard, the advantages and disadvantages of the block matrix obtained above, must be examined. The disadvantages will guide us on eliminating inefficient solution methods, while the advantages may help in constructing efficient ones.

In this section we consider the disadvantages.

The block matrix system in Fig. 6.5 can be written as

$$[A]\{X\} = \{B\} \tag{6.2}$$

where $\{X\} = \{u, v, p\}$ is the block solution vector containing the unknown velocities and pressures. To appreciate how huge this matrix may be, suppose we have a two dimensional problem involving, say, only 30 interior nodes in each of the two directions. In such a case there would be $30 \times 30 \times 3 = 2700$ unknowns, and A would be a $2{,}700 \times 2{,}700$ matrix with 7.29 million coefficients. Of these, however, less than 17,000 are nonzeros, so that only about one in each 450 coefficients is nonzero. The reader can imagine the case when, say, 1,000 nodes are chosen in each direction.

For a three dimensional problem, if, typically, we choose 30 points is the third direction, A would become a $81{,}000 \times 81{,}000$ matrix with 6,561,000,000 coefficients of which less than 702,000 are nonzeros. In such a case, less than one in every 9112 coefficients are nonzeros. This means that only 0.011 % of the matrix contains nonzero coefficients.

The above observation shows that the matrix A is extremely sparse, that is to say, the percentage of the zero terms is quite low.

It is clearly easy to conclude that not only the storage of this much of coefficients into a computer's memory is void, but also and most importantly, the solution of such a system is an extremely challenging problem.

There are, of course, various types of solution procedures that can be applied for the solution of general matrix systems. The utilization of a direct Gaussian elimination method that does not exploit the sparsity of the matrix A is completely impractical even for small sized problems. In the meantime the elimination process will bring nonzero coefficients to the places of the block matrix that were originally zeros. These extra nonzero terms will need extra storage and extra time for their re-elimination. Moreover, the execution time with such solvers will be very high. Therefore the utilization of such solvers for the present block system is quite out of the question.

To decrease the size of storage needed, one alternative may be to rearrange the equations so that A has a minimum bandwidth and to solve the set with a band solver. This may greatly reduce the storage and computing time since many of the nonzeros will no longer be included in the solution process. Still, the storage and time required for the solution may be prohibitive. Moreover, besides being complicated and requiring extensive memory space, an efficient and professionally written code for such a band solver may be out of reach of a humble researcher.

It is possible to obtain an alternative structure for the matrix A by a reordering of the equations within the block system. One way is to write the momentum and mass conservation equations in sequence, say, as for u_1, v_1, p_1, followed by u_2, v_2, p_2 etc. This produces a five-block-diagonal matrix, each of whose coefficients are actually 3×3 matrices. To obtain an idea on how the resulting matrix really looks like, the reader may try the creation on the example 3×3 grid used above.

This system can be solved iteratively or directly by treating the 3 × 3 submatrices as block coefficients of the new block matrix. Alternatively, the solution can be obtained by using a coupled direct-and-iterative solution approach. An iterative approach for the overall block system with 3 × 3 block submatrices, complemented by direct solution of the inversion process for the 3 × 3 submatrices may be considered. In such procedures, the zero terms appearing in the block 3 × 3 sub-matrices must be dealt with care. Attempts have been made in the past to apply such procedures. However, the experience of the Author and others show that the convergence rate of such procedures is low, and solution times are high (see Appendix A).

Turning back to the primitive form of the equations, let us consider the possibility of applying an iterative solution technique. In order that an iterative process works properly and convergently, the matrix of the system must possess two important properties. The matrix must not have any zero coefficients in its main diagonal and the matrix must be diagonally dominant. Diagonal dominancy is a condition saying that the magnitude of the main diagonal coefficient in each row of the coefficient matrix is greater than the sum of the magnitudes of the off-diagonal coefficients in that row. If the first condition is violated, the procedure will not work at all due to division by zero. If the second condition is violated, then the process will diverge. If the diagonal dominancy is weak, that is to say, if in most of the rows the magnitude of the main diagonal coefficient is equal to the magnitudes of the sums of the off-diagonal coefficients, then the iterative process may converge very slowly. In such a case the solution cost will be high.

In our case, direct application of an iterative solver is not possible because of the presence of zero terms in the main diagonal. Moreover, none of the rows of the block matrix are diagonally dominant. The third block row representing mass conservation is straightforwardly non-diagonally dominant. The first two row blocks too, are not diagonally dominant due to the coefficients in the submatrices A_{13} and A_{23}. Pivoting (i.e. interchanging the places of rows and columns) may help in transporting nonzero coefficients to the main diagonal. However, the penalty of this action will be a de-structuring of the original matrix, leading to a series of further problems. In such a case still, the diagonal dominance will not be achieved.

Researchers, in the past decades, considering the fact that there exists no 'actual equation' for pressure, worked out alternative procedures, called *segregated* solution procedures. All of these procedures, a summary of two of which is given in Appendix B, start by calculating a velocity field by assuming that a preliminary pressure field is given. This is followed by 'creating' a pressure equation by substituting the new velocity field into the mass conservation equation, but this time keeping the pressure terms in the momentum equation as variable unknowns.

Unluckily, however, the pressure field obtained from the solution of such 'pressure' equations in fact does not correlate with the velocity changes, so that it needs correction, which in turn necessitates the creation and solution of a 'pressure correction' equation of the same type as the pressure equation.

Since the equations obtained for this purpose, however, are of the Poisson-type (or elliptic), their solution is highly problematic. The problem is much more destructive in our case. The reason for this is that the pressure is 'free' at all boundaries, or mathematically speaking, all boundary conditions are of the form $\partial p/\partial n = 0$. At only one *reference* point in the solution region the value of pressure can be set to a fixed value, say zero. Therefore, all of the rows of the corresponding matrices for pressure and pressure correction will be weakly diagonally dominant, except the one at the reference point. This causes very slow convergence rates for an iterative solution process applied to these equations.

Although highly efficient solution procedures have been developed for the solution of the Poisson-type equations, the rate of convergence of the overall solution is still low.

It is very important to note at this stage that the 'segregation', thus imposed, effectively decouples the velocity and pressure variables. The strong coupling, however, was the most important property of the flow field, which should have been carefully preserved and perhaps utilized beneficially.

Despite their drawbacks, the segregated type of procedures had to be used in the past, perhaps since no alternative was formulated.

Then, there remains only one alternative for the solution of the primitive system. This is to find some way of bringing nonzeros to the main diagonal block submatrix of the mass conservation equations, but in the meantime provide, perhaps implicitly, diagonal dominancy to the resulting system and then resort to an iterative process. While this is done, the strong coupling between the velocity and pressure fields should be preserved.

The procedures which will be introduced in Chap. 7 constitute an up-to-date remedy for all of the above mentioned deficiencies.

Chapter 7
The Solution Procedure: Block Incomplete Decomposition

It was pointed out previously that looking at the governing differential equations themselves is misleading since it gives the impression that there is no equation for pressure. This illusion is due to the fact that there is no equation which contains dominant derivatives for pressure. It is not further possible to obtain such an equation by manipulating the governing differential equations.

Luckily enough, however, it is possible to obtain the pressure field implicitly, not by concentrating on the differential forms of the equations, but rather on the primitive form of their discretized counterpart. In fact, if we take a closer look at the lower block row and the rightmost block column of the block matrix formed in the last chapter, we see that, although implicit, in fact *there is* an equation for pressure. The only thing missing is some way of getting nonzero coefficients onto the main diagonal of the block lower right part of the primitive matrix. As we discussed before, there may be some way of achieving this, but the penalties are high.

The alternative to this is to divert attention from the primitive matrix and concentrate on some equivalent and useful partitioning of the matrix whose corresponding sub-matrices may contain nonzeros. This may even help, though indirectly, in gaining some kind of diagonal dominance as well.

On the other hand, it is important to note that actually the magnitudes of pressures are not so important for the velocity field. Rather, the pressure *differences* between the grid points are vital on how the velocity field will act. Therefore, instead of running after an explicit pressure field, it would be more meaningful to let the pressure differences act freely on their own to affect the velocity field, but in the meantime focus attention on conserving mass.

At this point, a 'tool' is required to achieve the above ideals. In fact, the 'innocent looking' mass conservation equation and its transposal part containing the effects of the pressure differences are the two main parts of this tool.

© Springer International Publishing Switzerland 2015

Z. Mazhar, *Fully Implicit, Coupled Procedures in Computational Fluid Dynamics*, Fluid Mechanics and Its Applications 115, DOI 10.1007/978-3-319-29895-5_7

The splitting technique, hereby named as the *incomplete decomposition process* that is described in the following sections provides an efficient mechanism for this purpose.

7.1 Properties and Advantages of the Block Matrix

In this section, instead of its disadvantages, we will concentrate on the useful properties of the block matrix system created in the previous chapter. After understanding the beneficial facts, it will be much easier to appreciate why and how the solution mechanism is formed.

Firstly, we note that the coefficients in each of the submatrices of the block matrix A are represented ready for use as vectors. There are five vectors for u and two vectors for pressure in the u momentum equations. Similarly, we have five vectors for v and two vectors for pressure in the v momentum equations. For mass conservation we have two vectors for u and two vectors for v. Therefore the complete block matrix can be cast into a total of eighteen vectors, each of dimension N. For a three dimensional problem this amounts to 24 vectors.

Some saving in memory can be obtained by not storing the vectors in the submatrices for pressure and mass conservation, since their coefficients may be calculated when the need arises. This is because they only contain the grid sizes of the control volumes.

Secondly, when we look at the last block row of the block matrix as a system with the right hand side vector, we can easily see that this last equation really acts on its own as an equation for pressure. This is true, despite the very unusual, but interesting fact that pressure itself does not appear in the equation. However, the right hand side vector of this row, b^p, contains the effects of the boundary conditions for the velocity field, which were enforced onto the mass conservation as well. From this, it is clear that the coefficients of the vector b^p, together with the mass conservation constraint itself, will affect the pressure field.

A question may be asked at this point: what about the boundary conditions for pressure? To answer this question, we note that the boundary conditions for pressure are imposed not onto the mass conservation equations, but rather, onto the two momentum conservation equations through their pressure terms as seen previously. Therefore, there are no means to worry about these conditions in regard of the mass conservation equation. The effects of the boundary conditions will be transferred through the implicitness of the cascaded process of the algorithm applied.

There exists, however, another issue to worry about. This is how we should take care of the relativeness of the pressures. It is surprising to note here that nothing needs to be done in this respect. In fact, pressures will set themselves implicitly while the solution progresses. Although the magnitudes of pressures may differ, the relative changes do not differ. This is due to the fact that only pressure differences appear in the momentum equations. However, since some high velocity changes may be expected especially at the early stages of the solution process with some inconvenient under-relaxation, the magnitudes of the pressures may tend to increase, mostly expected in the areas of the fluid where separation or swirling exists. This increase may not affect the solution, but may cause overflow in computer's memory, resulting in an unnecessary complete crash. Noting that some pressures will be negative and some will be positive, it will be enough, at some appropriate stage of the solution process, to set the pressure at a reference point, to some finite value (say 0). Subtracting a certain value from all of the values of pressures at all nodes is an alternative way of treatment. This will not make any effect on the momentum equations.

There remains a final, but most important problem to be addressed. For the solution of the pressure field, we need to have nonzero coefficients in the main diagonal sub-matrix A_{33} in Fig. 6.5. Although not beneficial, there is, in fact, one way of achieving this by directly working on the primitive matrix.

For this we may consider applying a standard elimination process on the system in Fig. 6.5 to eliminate A_{31} and A_{32} by using the two momentum equations. This provides an explicit equation for pressure as $-[A_{13}A_{11}^{-1}A_{31} + A_{23}A_{22}^{-1}A_{32}]p = b^p - b^u A_{11}^{-1}A_{31} - b^v A_{22}^{-1}A_{32}$. Now the pressure field can be obtained by solving this system. After the pressure field is thus obtained, a back-substitution process then will yield the velocity components. The pressure and velocity fields, thus found surely will satisfy mass and momentum conservations simultaneously, of course with the provided u^* and v^* values. The whole process, then, needs to be repeated using the new velocity field in order to treat the nonlinearities in the momentum equations.

Mathematically everything seems perfect with this process. However, although the above remedy seems adequate and appealing, we need to answer the following questions before attempting such a solution: How are we going to obtain the inverses of the main diagonal block sub-matrices? Moreover, what will the structure of the matrix in parentheses be? These are, for sure, challenging questions to answer. In fact, this will be like, with an old saying, 'facing with a hail squall while running away from rain'. By embarking on such an adventure, eventually we will fall into a trap similar to the ones in the segregated procedures.

One way out of this dilemma may be to transfer all of the off-diagonal coefficients in the main block diagonal matrices of the momentum equations to the right hand side vectors of the system and later resort to an iterative sequence. The inversion of the updated block matrices $A_{11}^{'}$ and $A_{22}^{'}$ is now easy since they contain only a main diagonal vector. Once that the problem with the inversions is eased in this way, the linear system constructed for pressure will be of the usual Poisson-type. This is still a serious problem since the matrix on the left hand side will be loosely diagonally dominant, a property which leads to slow convergences in an iterative solution process. For this reason, and because we have set forward for getting rid of a Poisson-type equation, the above alternative will not be pursued any further.

If we can afford the solution of a Poisson-type equation, the performance of such a procedure will surely be much better than that of a segregated procedure. This is because not much harmful assumptions have been made in due course and that mass and momentum conservation will be simultaneously satisfied at the end of each block solution.

It would be most appreciative if we can find some way of solving the primitive equations without the need for solving a Poisson-type equation. For this, in the sections that follow, we will resort to an alternative, simple, yet extremely powerful strategy, which provides the solution of the coupled system in its primitive, original form.

Nevertheless, the above discussion provides an extremely important observation which should not be missed. By looking at the expression in parenthesis of the above equation, we can immediately realize an equivalent symmetric effect of the mass conservation and pressure terms through the submatrices A_{13}, A_{31} and A_{23}, A_{32} which multiply the inverse momentum matrices on the left and right. This observation clearly asserts our claim in Sect. 3.3 that the pressure effects are as important as those of the mass conservation principle.

The mechanism involved in the procedures presented in the following sections explores this physically crucial and critical property of the flow of incompressible fluids in an efficient way.

7.2 General Incomplete Decomposition

Let us be given a general linear system of equations denoted by

$$AX = B \tag{7.1}$$

The solution of such a system by a classical method may not be feasible, especially when the matrix A is sparse. It may be possible at times that the matrix A can be easily decomposable into two submatrices L and U such that

$$A = LU \tag{7.2}$$

In this case, Eq. (7.1) can be written as

$$LUX = B \tag{7.3}$$

Now letting

$$UX = Y \tag{7.4}$$

reduces Eq. (7.3) to

$$LY = B \tag{7.5}$$

Then the solution of the system (7.1) can be accomplished in two steps:

Step 1: Solve $LY = B$ for the intermediate vector Y
Step 2: Solve $UX = Y$ for the unknown vector X

The benefit of this type of an approach is twofold. Firstly, if the forms of the matrices L and U are adequately selected, the solutions in the above two steps may be easy and fast. Secondly, the decomposition process may need to be done only once for a complete problem, if many systems need to be solved which involve the same matrix.

If the matrices L and U are strictly lower and upper triangular, the procedure is named as the LU decomposition, and is very well known in the literature. In this case, the two lower and upper triangular systems are easy to solve, requiring a forward and a backward substitution (or a linear sweep), respectively.

Although not usual, however, the two matrices L and U may not be lower or upper triangular, but of some other suitable form. The only criterion in the choice of how their structure will be is that the solution of the corresponding two systems be straightforward. For example, a tri-diagonal or similar form may even be much appreciated, if possible.

The above process constitutes a direct solution to the given system, which, of course is the most appreciative. However, in some cases, a complete decomposition of the above type may not be possible. Even if it may be possible, the structure of the two matrices may be such that too many extra nonzero coefficients may appear in the places which contain zeros in the original matrix. Thus, the solution of the two systems may not be efficient due to high storage and time requirements. In such a case an alternative is to use an 'incomplete' decomposition, details of which we now present.

Let A be decomposed as

$$A = LU + D \tag{7.6}$$

in which D will hereby be named as the *defect matrix* and the L and U matrices as the *strength matrices.* The vectors of the defect matrix will be called the *defect vectors* and the vectors of the strength matrices the *strength vectors.*

Then the system in Eq. (7.1) can be written as

$$(LU + D)X = B \tag{7.7}$$

There are two equivalent methods for obtaining the solution of this system with an iterative process, each of which has its own advantages and disadvantages:

Method 1
Write the Eq. (7.7) as

$$LUX = B - DX \tag{7.8}$$

The solution of the original system (7.1) can now be accomplished by the successive iteration

$$LUX^{(n+1)} = B - DX^{(n)} \tag{7.9}$$

Then, the solution can be obtained in three steps as follows:

Step 1: Solve $LY^{(n+1)} = B - DX^{(n+1)}$ for $Y^{(n+1)}$
Step 2: Solve $UZ^{(n+1)} = Y^{(n+1)}$ for $Z^{(n+1)}$
Step 3: Update X by $X^{(n+1)} = (1-\alpha)X^{(n+1)} + \alpha Z^{(n+1)}$

The iteration thus formed starts with a preliminary assumption $X^{(0)}$ and continues until satisfactory convergence is achieved. The parameter α is a relaxation factor which may have a value larger than 0 and less than or equal to 1.

Method 2
For this method we first subtract $LUX^{(n)}$ from both sides of Eq. (7.9) to get

$$LUX^{(n+1)} - LUX^{(n)} = B - DX^{(n)} - LUX^{(n)} \tag{7.10}$$

from which we can write

$$LU\left(X^{(n+1)} - X^{(n)}\right) = B - (LU + D)X^{(n)} \tag{7.11}$$

Replacing $LU + D$ by A and letting $\delta^{n+1} = X^{(n+1)} - X^{(n)}$ we get

$$LU\delta^{(n+1)} = B - AX^{(n)} \tag{7.12}$$

Now, one iteration of the solution process consists of the following steps:

Step 1: Solve $LY^{(n+1)} = B - AX^{(n+1)}$ for $Y^{(n+1)}$
Step 2: Solve $U\delta^{(n+1)} = Y^{(n+1)}$ for $\delta^{(n+1)}$
Step 3: Update X by $X^{(n+1)} = X^{(n)} + \alpha\delta^{(n+1)}$

The two methods are equivalent and they produce the same results. To decide on which method to prefer depends on the number of operations performed to calculate the right hand side in Step 1. Normally Method 1 would be preferred since the number of defect vectors in D is expected to be less than those in the matrix A. Consequently, the number of operations to calculate the right hand side in Step 1 of Method 1 will be much less than that of Method 2.

On the other hand, it can be observed that the right hand side of the system in Step 1 of Method 2 is the *residual* of the original system. The residual is an indication of the level of convergence of the iterations. In our case, the level of convergence also characterizes the simultaneous satisfaction of the mass and momentum conservations, which is physically more explanatory. Therefore, if the residual is calculated for convergence checking, it might be readily used in the solution as well. This would require no extra effort. In this case, the use of Method 2 is more feasible.

There is one more important advantage of Method 2. In this method, the calculation of the residuals in Step 1 does not involve the utilization of the defect vectors that will be constructed in the special procedures devised for our special block matrix system. Since the defect vectors are not needed elsewhere, they need not be calculated or stored at all. This is an important saving in memory usage and computational effort.

For these reasons, the use of Method 2 is advisable and will be utilized in what follows.

The right hand side of the system in step 1 of Method 1 is not the residual of the block system. Therefore, with Method 1, some other criteria may be used for convergence testing. A good practice is to use the *maximum pressure differences* test given by

$$\max_i \left| p_i^{n+1} - p_i^n \right| < \varepsilon \tag{7.13}$$

for some suitable small value of ε.

Before commencing with the application of the above process to our special primitive form, we note that the only necessary and sufficient condition for the process to converge is

$$\left\| (LU)^{-1} D \right\|_2 < 1 \tag{7.14}$$

In plain words, this condition states that if the effect of the defect matrix D is less than the effect of the product LU, then the process converges.

7.3 An Incomplete Decomposition of the Block System (BIP)

Once that we have decided on the main general strategy, it is the time to delve into the details of the incomplete decomposition process that is suitable for our block coupled system. In fact, there exist various ways of doing this, each of which leads to a different procedure.

The main decision here is on the choice of the shapes of the strength matrices L and U. The shape of the resulting defect matrix then is established by this choice.

We now give a complete detail of the incomplete decomposition process for our block matrix system composed in the previous chapter, by considering one of the possibilities for the shapes of L and D. In later sections, we present some other choices as well.

Consider taking the L, U and D matrices as shown in Figs. 7.1, 7.2 and 7.3, respectively.

In order to see the relationship between the coefficients of A and those of L, U and D, it is necessary to multiply the matrices L and U, add D, and equate the result to A. This operation results in the following equations

$$A^p_{W_i} = A^u_{P_i} h_i, \quad i = 1, N \tag{7.15a}$$

$$A^p_{E_i} = A^u_{P_i} f_i, \quad i = 1, N-1 \tag{7.15b}$$

$$A^p_{S_i} = A^v_{P_i} k_i, \quad i = 1, N \tag{7.15c}$$

$$A^p_{N_i} = A^v_{P_i} n_i, \quad i = 1, N-m \tag{7.15d}$$

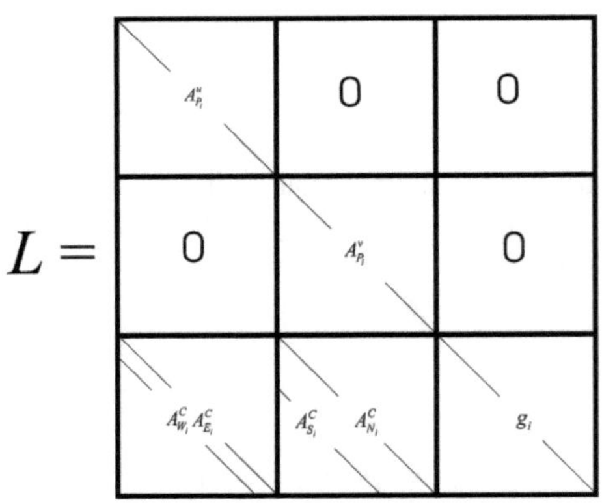

Fig. 7.1 Block strength matrix L for BIP

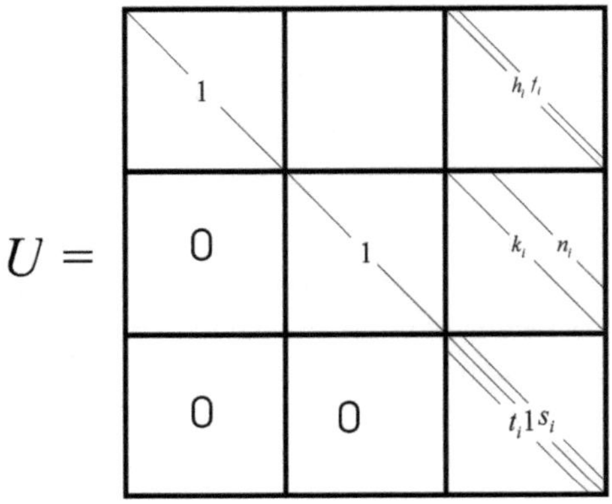

Fig. 7.2 Block strength matrix U for BIP

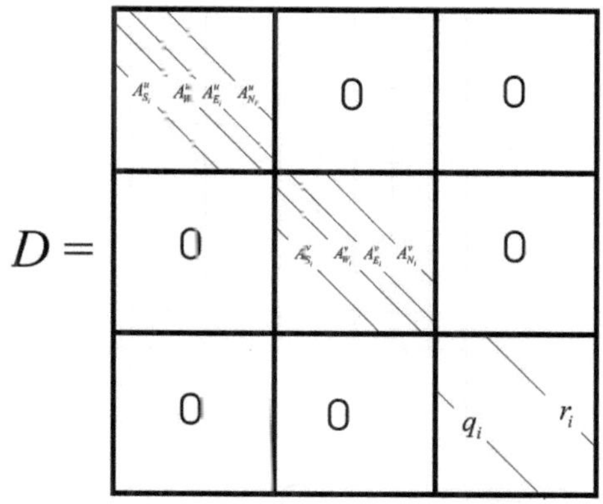

Fig. 7.3 Block defect matrix D for BIP

$$A^C_{W_i} f_{i-1} + A^C_{E_i} h_i + A^C_{S_i} n_{i-m} + A^C_{N_i} k_i + g_i = 0, \quad i = 1, N \tag{7.15e}$$

$$A^C_{E_i} f_i + s_i g_i = 0, \quad i = 1, N - 1 \tag{7.15f}$$

$$A^C_{W_i} h_{i-1} + t_i g_i = 0, \quad i = 2, N \tag{7.15g}$$

$$A^C_{S_i} k_{i-m} + q_i = 0, \quad i = 1 + m, N \tag{7.15h}$$

$$A^C_{N_i} n_i + r_i = 0, \quad i = 1, N - m \tag{7.15i}$$

At a first glance, it is not really obvious to realize how these equations are constructed. It is advisable at this point that the reader uses the matrix obtained for the 3 × 3 grid as given in Fig. 6.4, then fully construct the corresponding matrices L, U and D by using the shapes given in Figs. 7.1, 7.2 and 7.3. The full shapes are given in Figs. 7.4, 7.5 and 7.6.

The construction of the Eqs. (7.15a–7.15d) is straightforward and needs no further elaboration. For the rest of the equations, the highlighted 5-th row of the matrix L in Fig. 7.4 in the third row-block is taken and turned clockwise by 90 degrees, then multiplied coefficient by coefficient by each column of the third block-column of the matrix U in Fig. 7.5. After obtaining the equations for this 5th row, the general equations for the coefficients in the ith row can be obtained by replacing the subscripts denoted by 5 by i, so that 2's become $i - m$ and 4's become $i - 1$, noting that $m = 3$.

It will be much easier for the reader to work with a 4 × 4 or better with a 5 × 5 system and using the 7th block-row or the 13th block-row, respecticely, of $[L_{31} \quad L_{32} \quad L_{33}]$. This will require four large 48 × 48 or 75 × 75 squared papers put together side by side for the matrices A, L, U and D. We leave this to the enthusiastic reader, since the dimensions of the pages of the bock are too limited for such a representation. However, for the construction of the more complicated versions

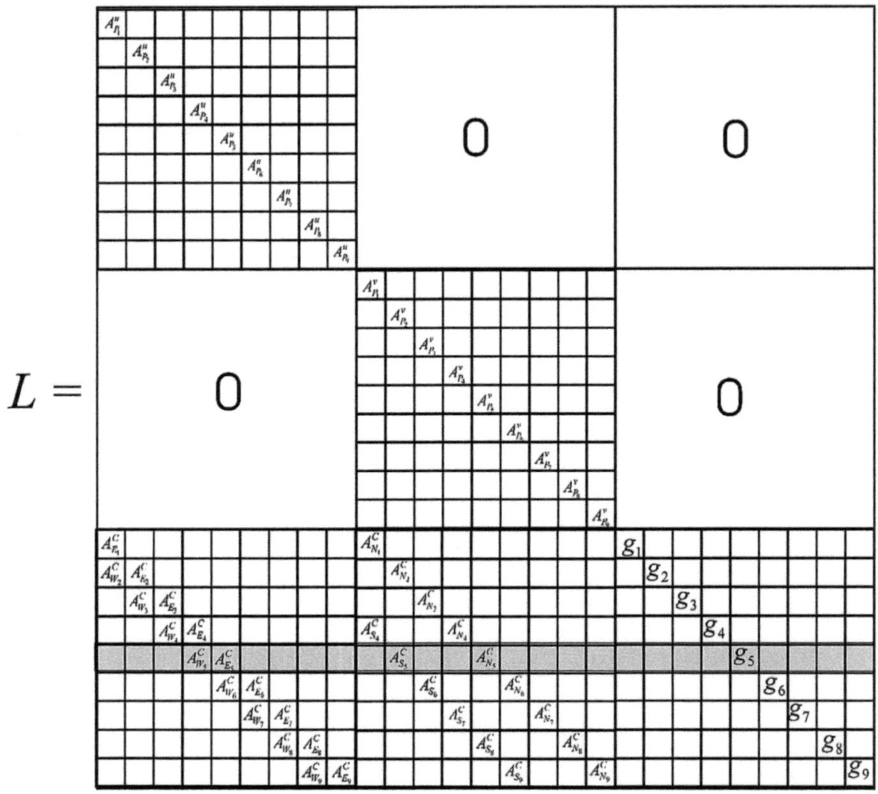

Fig. 7.4 Full form of the block strength matrix L for BIP

that we will present later on, the utilization of a 75 × 75 case is advisable for a full appreciation of the construction of the formulations.

Note: The novice reader may initially choose to work the process with an easier version by setting t_i and s_i to zero. This, in fact, will lead to a version which will be given special attention later on.

Now the coefficients of the nine vectors, h_i, f_i, k_i, n_i, g_i, t_i, s_i, q_i and r_i can be computed from Eqs. (7.15a–7.15i) as follows:

$$h_i = A^p_{W_i} \Big/ A^u_{P_i}, \quad i = 1, N \tag{7.16a}$$

$$f_i = A^p_{E_i} \Big/ A^u_{P_i}, \quad i = 1, N-1 \tag{7.16b}$$

$$k_i = A^p_{S_i} \Big/ A^v_{P_i}, \quad i = 1, N \tag{7.16c}$$

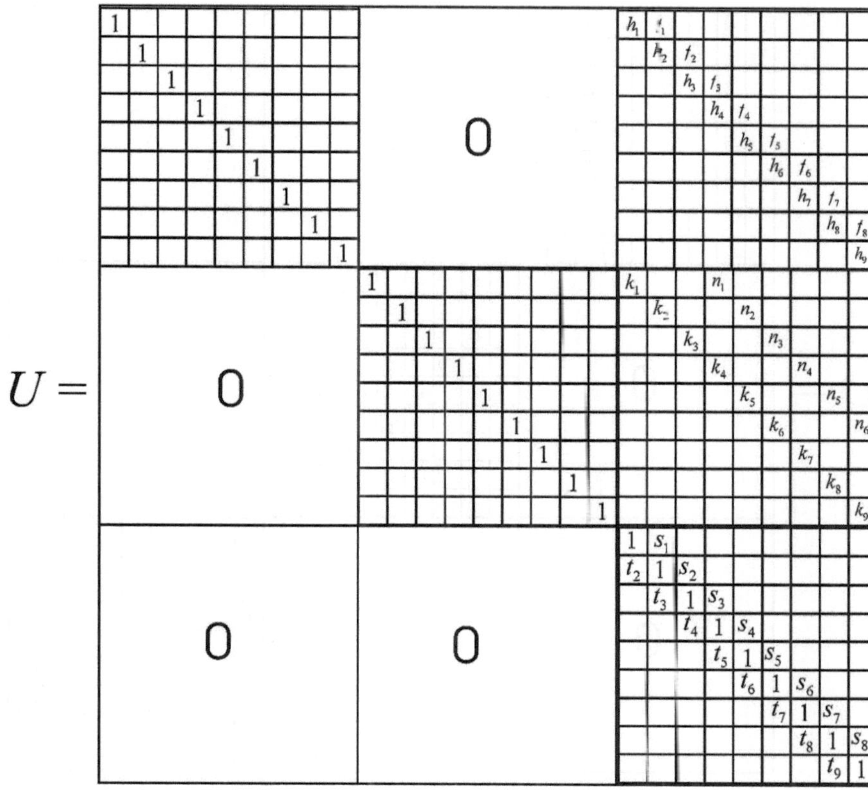

Fig. 7.5 Full form of the block strength matrix U for BIP

$$n_i = A_{N_i}^p \Big/ A_{P_i}^v, \quad i = 1, N - m \tag{7.16d}$$

$$g_i = -\left(A_{W_i}^C f_{i-1} + A_{E_i}^C h_i + A_{S_i}^C n_{i-m} + A_{N_i}^C k_i\right), \quad i = 1, N \tag{7.16e}$$

$$s_i = -A_{E_i}^C f_i \Big/ g_i, \quad i = 1, N - 1 \tag{7.16f}$$

$$t_i = -A_{W_i}^C h_{i-1} \Big/ g_i, \quad i = 2, N \tag{7.16g}$$

$$q_i = -A_{S_i}^C k_{i-m}, \quad i = 1 + m, N \tag{7.16h}$$

$$r_i = -A_{N_i}^C n_i, \quad i = 1, N - m \tag{7.16i}$$

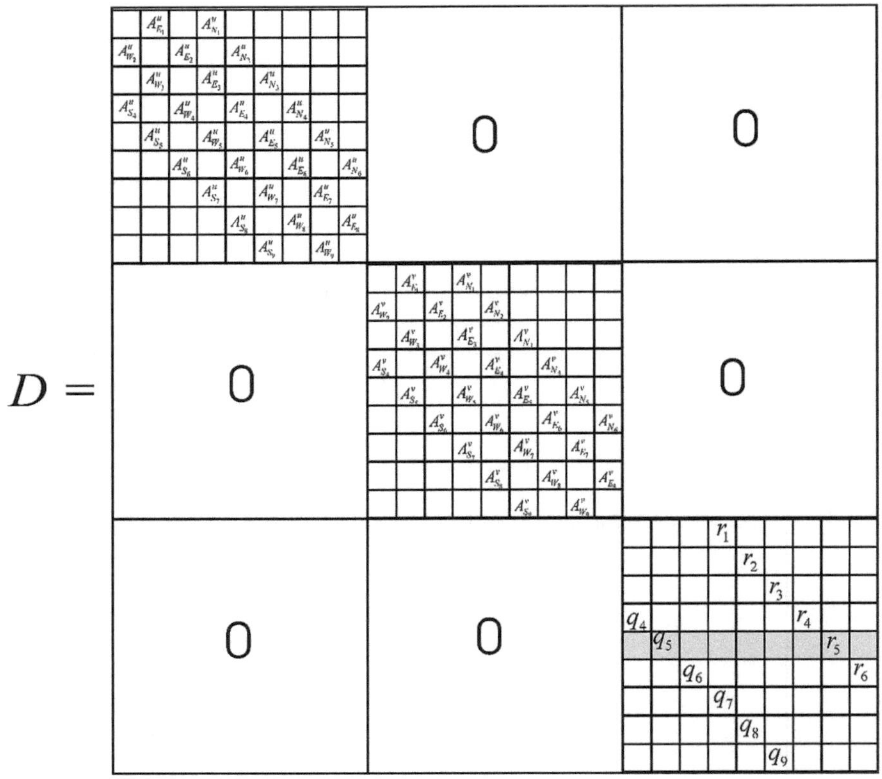

Fig. 7.6 Full form of the block defect matrix D for BIP

At this point, a brief discussion of the above equations is essential:

1. The coefficients h_i, f_i, k_i and n_i can be computed directly, without any problem.
2. The coefficients g_i contain coefficients h_i, f_i, k_i and n_i with varying subscripts, and hence for some subscripts they will not be defined. For example f_{i-1} is not defined for $i = 1$, and n_{i-m} is not defined for $i = 1, m$. Therefore, g_i must be calculated separately for this range of the subscripts.
3. The coefficients s_i, t_i, q_i and r_i must be calculated after all the coefficients h_i, f_i, k_i, n_i and g_i are calculated. When Method 2 is used for the solution, the coefficients q_i and r_i need not be calculated.
4. With a properly formed primitive matrix, the coefficients g_i will not become zero, hence the solution of the system in Step 1 will not produce any problem. In addition, the tri-diagonal solution of the system in Step 2 will not produce any zeros on the main diagonal of U.

A discussion of the mechanism experienced with these equations is left to Sect. 8.3, until after applications are presented.

7.4 The Block Solution Procedure

Now that the decomposition has been performed, the following steps are followed to complete one iteration of the solution process:

1. Calculate the 'residual vector' $R^{(n)} = B - AX^{(n)}$, using the already calculated values of $X^{(n)} = \left(u^{(n)}, v^{(n)}, p^{(n)}\right)$, or use some initial values, when $n = 0$.
2. Solve the system $LY^{(n+\frac{1}{2})} = R^{(n)}$ to obtain the vector $Y^{(n+\frac{1}{2})} = \left(u^{(n+\frac{1}{2})}, v^{(n+\frac{1}{2})}, p^{(n+\frac{1}{2})}\right)$. This is straightforward and requires a forward solution in three steps:

   a. Solve $u_i^{(n+\frac{1}{2})}$ by using $u_i^{(n+\frac{1}{2})} = R_i^{u^{(n)}} \Big/ A_{P_i}^u$
   b. Solve $v_i^{(n+\frac{1}{2})}$ by using $v_i^{(n+\frac{1}{2})} = R_i^{v^{(n)}} \Big/ A_{P_i}^v$
   c. Solve $p_i^{(n+\frac{1}{2})}$ by using $p_i^{(n+\frac{1}{2})} = \left[R_i^{p^{(n)}} - \left(A_{W_i}^C u_{i-1}^{(n)} + A_{E_i}^C u_i^{(n)} + A_{S_i}^C v_{i-m}^{(n)} + A_{N_i}^C v_i^{(n)}\right)\right] \Big/ g_i$

3. Solve the system $U\delta^{(n+1)} = Y^{(n+\frac{1}{2})}$ to obtain the vector $\delta^{(n+1)} = \left(\delta_u^{(n+1)}, \delta_v^{(n+1)}, \delta_p^{(n+1)}\right)$. This requires a backward solution in three steps:

   a. Solve the tridiagonal system by a forward and backward process for calculating the velocity and pressure differences

      i. Eliminate t_i by a forward elimination
      ii. Set $\delta_{p_N}^{(n+1)} = 0$
      iii. Solve backward to obtain $\delta_p^{(n+1)}$ for $i = N - 1$ down to 1.

   b. Solve for $\delta_v^{(n+1)}$ by backward substitution
   c. Solve for $\delta_u^{(n+1)}$ by backward substitution

4. Update X by $X^{(n+1)} = X^{(n)} + \alpha\delta^{(n+1)}$ to get $X^{(n+1)} = \left(u^{(n+1)}, v^{(n+1)}, p^{(n+1)}\right)$

Note that Step 3a(ii) is to cure the relativeness of the pressures and to avoid overflow in the process of solution.

The above steps are repeated for $n = 1, 2, \ldots$, until a satisfactory level of convergence of the approximations is established. Complete convergence is achieved when the residual vector denoted by $R^{(n)} = B - AX^{(n)}$ in Step 1 becomes zero, or becomes close to zero within machine accuracy.

When convergence is achieved, the simultaneous satisfaction of the mass and momentum conservations is guaranteed within the block solution. However, it is not really necessary to bring the residuals to zero in the block solutions, since the results at the end of a block solution will not be the actual solution of the flow field. This is due to the nonlinearities in the governing differential equations, discussed previously.

7.5 Complete Solution of the Flow Field

For the complete solution of the flow field, it is necessary to recalculate the coefficients in the discretization equations and repeat the solution process described in Sect. 7.4 above. This is required for the treatment of the nonlinearities resulting from the first order derivative terms appearing in the momentum equations. A full iteration process consisting of

1. Forming a new block matrix system by refreshing the coefficients using the recently calculated velocity field
2. Applying block implicit solution

will be named as an 'overall iteration'.

The convergence of the overall iterations to the actual flow field, then, may be described by a condition, assuring that the changes in the flow field from one overall iteration to another are small enough, or negligible.

A suitable criterion for this is the maximum absolute difference in the pressure field between two overall iterations, normalized by the head of the 'entering' fluid as

$$\varepsilon_p = \frac{\max_i \left| p_i^{s+1} - p_i^s \right|}{\frac{1}{2}\rho u_{ref}^2} \tag{7.17}$$

in which u_{ref} is some fixed velocity value given at some reference point in the solution region. This, usually, is the value of the velocity at an entering point of the solution region.

7.6 A Family of Procedures: BIPEN, FICS-1, FICS-2

A question which the reader may have asked for himself at this point is why the strength vectors in the L and U matrices are so placed, and isn't there other ways of selection, or places that other vectors may be employed in these matrices? Furthermore, how are the places of the vectors in the defect matrix selected?

We first answer the latter question. The vectors in the defect matrix, and the places at which they are placed are not actually decided beforehand, but rather established by the places of the strength vectors in the L and U matrices. The reader may have already gained an insight into this subject while working on the construction of the formulas.

The following considerations may be useful in the selection of these vectors:

1. Less number of vectors means less extra storage and computing time. More number of strength vectors may give more durable and faster convergence, provided that extra storage is tolerated.
2. Less number of defect vectors may mean smaller defective effects leading to a faster convergence of the solution process.
3. For some selections of the strength vectors, impossibilities may arise. This, for example, may occur if any strength vector is chosen at the upper part of L, and at the much lower part of U.

To answer to the former issue; yes, although not much, there are of course a variety of ways and places that the strength vectors in the L and U matrices can be placed.

All other sub-matrices remaining the same as described in Sect. 7.3, we now consider some examples of other possible ways of placement of the strength and defect vectors by taking the L_{33}, U_{33} and D_{33} sub-matrices as shown in Figs. 7.7, 7.8 and 7.9.

The vectors shown in these matrices will not all need be used at the same time. They are represented here for the sake of general reference. The choice of these vectors lead to different solution procedures. A list of some possible choices is given in Table 7.1. The places of the coefficients of the vectors in the corresponding diagonal row i of the respective submatrices are shown in the last row of the table.

Fig. 7.7 Possible choices for L_{33}

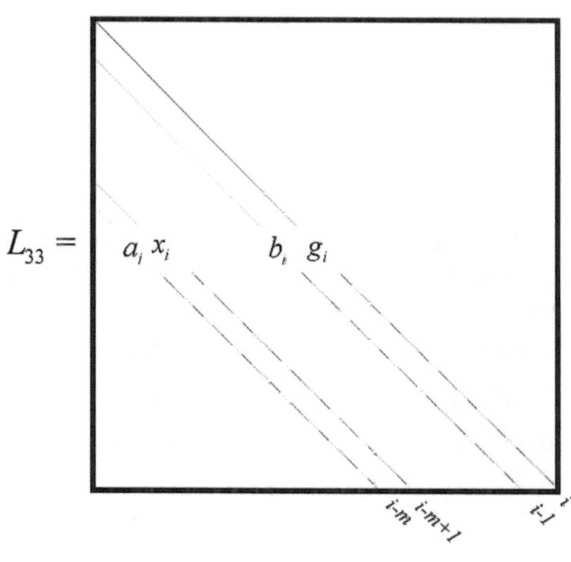

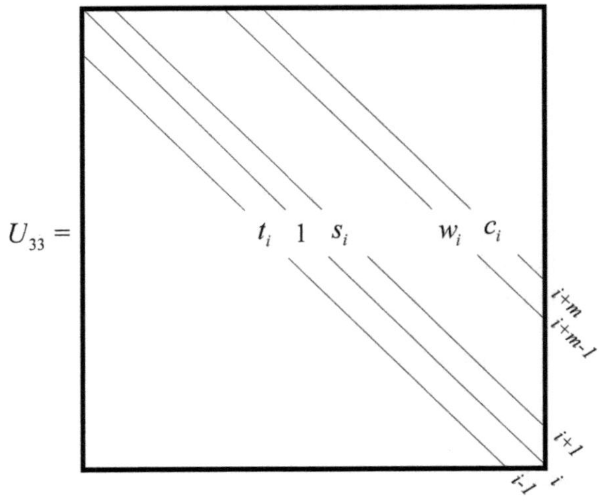

Fig. 7.8 Possible choices for U_{33}

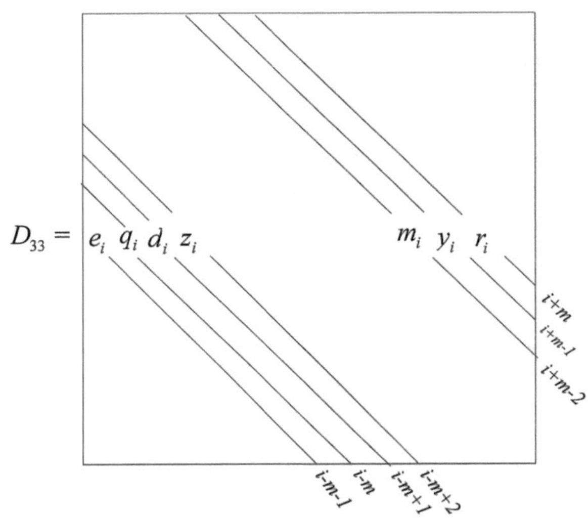

Fig. 7.9 Possible choices for D_{33}

The choice given in the row denoted by BIP leads to the method that was described in the previous section. BIPEN, FICS-1 and FICS-2 are enhanced versions, so named in the literature (see Appendix A).

Letting $G_i = A^C_{W_i} f_{i-1} + A^C_{E_i} h_i + A^C_{S_i} n_{i-m} + A^C_{N_i} k_i$, $h_i = A^p_{W_i} \big/ A^u_{P_i}$, $f_i = A^p_{E_i} \big/ A^u_{P_i}$ $k_i = A^p_{S_i} \big/ A^v_{P_i}$ and $n_i = A^p_{N_i} \big/ A^v_{P_i}$, the formulas for the strength and defect vectors for BIPEN, FICS-1 and FICS-2 are given as follows:

BIPEN	*Strength vectors*	*Defect vectors*
	$a_i = -A^C_{S_i} k_{i-m}$	$e_i = -a_i t_{i-m}$
	$g_i = -G_i$	$d_i = -a_i s_{i-m}$
	$t_i = -A^C_{W_i} h_{i-1} \Big/ g_i$	$r_i = -A^C_{N_i} n_i$
	$s_i = -A^C_{E_i} f_i \Big/ g_i$	
FICS-1	*Strength vectors*	*Defect vectors*
	$a_i = -A^C_{S_i} k_{i-m}$	$d_i = -a_i s_{i-m}$
	$b_i = -A^C_{W_i} h_{i-1}$	$y_i = -b_i c_{i-1}$
	$g_i = -G_i - a_i c_{i-m} - b_i s_{i-1}$	
	$s_i = -A^C_{E_i} f_i \Big/ g_i$	
	$c_i = -A^C_{N_i} n_i \Big/ g_i$	
FICS-2	*Strength vectors*	*Defect vectors*
	$a_i = -A^C_{S_i} k_{i-m}$	$z_i = -x_i s_{i-m+1}$
	$x_i = -a_i s_{i-m}$	$m_i = -b_i w_{i-1}$
	$b_i = -A^C_{W_i} h_{i-1} - a_i w_{i-m}$	
	$g_i = -G_i - a_i c_{i-m} - x_i w_{i-m+1} - b_i s_{i-1}$	
	$s_i = -\left(A^C_{E_i} f_i + x_i c_{i-m+1}\right) \Big/ g_i$	
	$w_i = -b_i c_{i-1} / g_i$	
	$c_i = -A^C_{N_i} n_i \Big/ g_i$	

Running indices for the strength and defect vectors are as follows:

Strength vectors		
$a_i : m+2, N$	$x_i : i = m-1, N$	$b_i : i = 2, N$
$g_i : i = 1, N$	$t_i : i = 2, N$	$s_i : i = 1, N-1$
$w_i : i = 1, N-m+1$	$c_i : i = 1, N-m$	
Defect vectors		
$e_i : i = m+2, N$	$q_i : i = m+1, N$	$d_i : i = m, N$
$y_i : i = 1, N-m$	$r_i : i = 1, N-m-1$	$z_i : i = m-1, N$
$m_i : i = 1, N-m+1$		

Table 7.1 Various choices for the strength and defect vectors

	Strength vectors								Defect vectors						
	L				*U*				*D*						
	a	*x*	*b*	*g*	*t*	*s*	*w*	*c*	*e*	*q*	*d*	*z*	*m*	*y*	*r*
BIP				•	•	•				•					•
BIPEN	•			•	•	•			•		•				•
FICS-1	•		•	•		•		•			•			•	
FICS-2	•	•	•	•		•	•	•				•	•		
Location in i-th row	*i-m*	*i-m+1*	*i-1*	*i*	*i-1*	*i+1*	*i+m-1*	*i+m*	*i-m-1*	*i-m*	*i-m+1*	*i-m+2*	*i+m-2*	*i+m-1*	*i+m*

7.7 Storage Requirements and Complexity

When a new procedure is forwarded, among other things, the researcher is interested basically on how costly the procedure is. The cost of a procedure mainly depends on the storage that would be required to store the extra vectors and on the total time spent for a complete solution. These, in turn, not only depend on the procedure itself, but also on the implementation and the programming technique used. These issues are discussed in this section.

7.7.1 Storage Requirements

To determine the storage requirements of the procedures, we must first determine the minimum (or necessary) requirements. Following this we must concentrate on how much storage we might save by certain clever arrangements.

By examining the primitive block matrix system, it is easy to conclude that storage for the vectors in A11 and A22 is compulsory. The three vectors on the right hand sides must also be stored. The vectors in A31, A32, A13 and A23 may, or may not be stored. The coefficients of these vectors contain only the grid sizes. Therefore, they may not be stored, but calculated when need arises in the calculation of the strength vectors. However, these vectors contain the effects of the boundary conditions on velocity and pressure. So, if it is decided that storage will not be used for these vectors, special precautions in programming must be taken while the boundary conditions are implemented.

The strength vectors in L33 and U33 need storage. Storage for three, four, five and seven of these vectors for the BIP, BIPEN, FICS-1 and FICS-2, respectively, is required. The coefficients of the vectors h, k, f and n need not be stored, since they may be calculated directly within the formulas of the strength and defect vectors.

If Method 2 is used in the block solutions, none of the defect vectors need storage, since they are not used. We note that, only the residual vector is used to calculate the right hand side of the lower triangular system.

Considering the above observations, with a professional and clever programming strategy, only a handful of vectors need to be employed in the solution procedures. Nevertheless, a novice user is advised hereby to store all the vectors in the primitive matrix and the strength matrices, at least until he is fully confident of the proper functioning of the algorithm.

7.7.2 *Complexity*

'Complexity' is a notational term used by computer scientists to define the number of operations performed in an algorithm. In our case, it represents mainly the total number of multiplicative operations, i.e., multiplications and divisions. If the algorithm gives the solution in a fixed, specified number of operations, the complexity would be used as an indication of the speed of the algorithm.

The complexity, as described above, is not solely an indication on the speed of a procedure used for the solution of fluid flow problems. This is because the number of (block) inner iterations and the number of overall iterations that need to be performed are not known apriori. All of these depend on the nature of the physical problem and the ability of the solution procedure to cope with the problem. For example, in large problems, the effects of the changes at some grid point in the solution region may not be appreciated quickly at all points, leading to slower convergence. Convergence may also be slowed down if a non-optimized relaxation parameter is used. These factors may increase the total number of overall iterations, as well as the operations performed in the inner iterations.

However, since most of the time in the solution process is spent in the inner iterations, it is useful to have an idea of the complexity of the algorithms used in the block solutions. Table 7.2 gives the number of multiplicative operations ($\times N$)

Table 7.2 Complexity of the algorithms

	BIP	BIPEN	FICS-1	FICS-2	
Coefficient calculation (only once)	12	13	16	22	Strength
	2	3	2	2	Defect
Solution (per iteration)	14	15	13	15	

performed to calculate the coefficients of the strength and defect vectors and the operations performed in the solution stages, for each procedure.

In this table, multiplications by a constant parameter are not included. Also not included is the number of operations performed for residual calculation.

We also note that the coefficient calculations are done once in a complete block solution. This is followed by a series of residual calculations, forward sweeps and backward substitutions for the solution. The number of preparatory operations performed elsewhere is not of much importance.

7.8 The Simplest Case (Simple Implicit Coupled Solution—SICS)

This section is intentionally devoted to introducing a procedure which has some very special properties. These properties, together with the benefits that are supplied with, are discussed at the end of the section, after the derivations are formulated.

First let the strength matrices be of the form

$L =$ 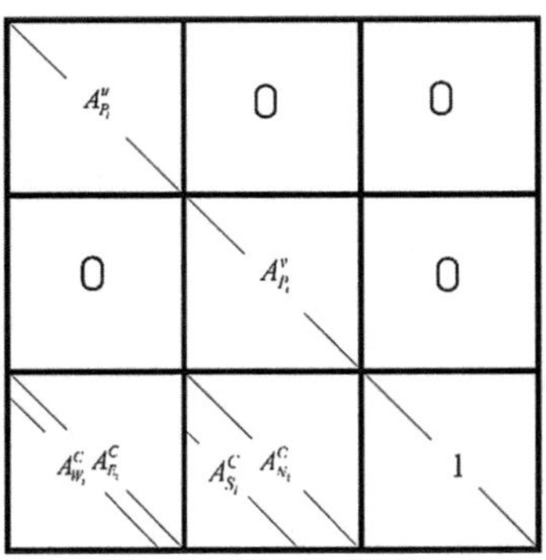

$U =$ 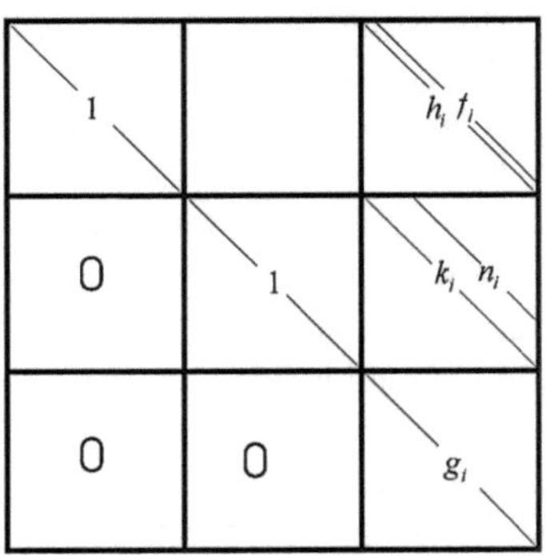

With this choice of the strength matrices, the shape of the defect matrix becomes as follows:

$D =$ 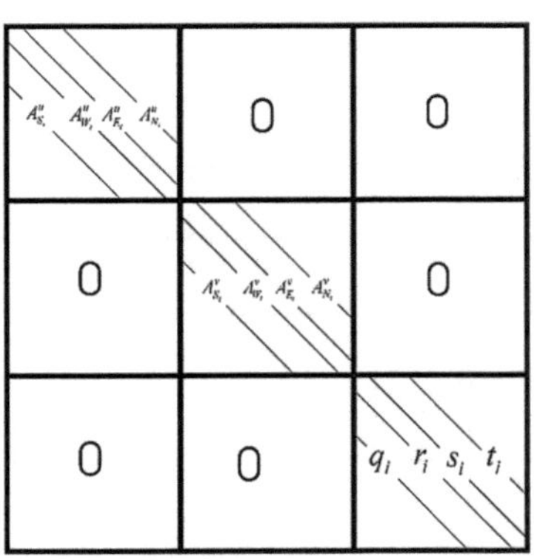

Now we multiply L and U, add D and equate to A to obtain the formulas for the coefficients of the strength vectors h, f, k, n and g, and those for the defect vectors q, r, s and t:

$$h_i = A^p_{W_i} \Big/ A^u_{P_i}, \quad i = 1, N \tag{7.18}$$

$$f_i = A^p_{E_i} \Big/ A^u_{P_i}, \quad i = 1, N - 1 \tag{7.19}$$

$$k_i = A^p_{S_i} \Big/ A^v_{P_i}, \quad i = 1, N \tag{7.20}$$

$$n_i = A^p_{N_i} \Big/ A^v_{P_i}, \quad i = 1, N - m \tag{7.21}$$

$$g_i = -\left[A^C_{W_i} f_i + A^C_{E_i} h_i + A^C_{S_i} n_i + A^C_{N_i} k_i\right], \quad i = 1, N \tag{7.22}$$

$$q_i = -A^C_{S_i} A^p_{S_{i-m}}, \quad i = m + 1, N \tag{7.23}$$

$$r_i = -A^C_{W_i} A^p_{W_{i-1}}, \quad i = 2, N \tag{7.24}$$

$$s_i = -A^C_{E_i} A^p_{E_i}, \quad i = 1, N - 1 \tag{7.25}$$

$$t_i = -A^C_{N_i} A^p_{N_i}, \quad i = 1, N - m \tag{7.26}$$

Although the defect vectors will not be employed in the solution process, they are given here for the sake of completeness. The locations of these coefficients in the ith row of D_{33} are shown below for easy reference.

q_i	r_i	s_i	t_i
$i - m$	$i - 1$	$i + 1$	$i + m$

7.8.1 The Solution Algorithm (SICS)

The algorithm of the above solution process is straightforward to the experienced reader. To aid the novice reader and to demonstrate its simplicity, we provide a complete description below.

We let

$$X^{(n)} = \left(u^{(n)}, v^{(n)}, p^{(n)}\right),$$
$$Y^{(n)} = \left(\hat{u}^{(n)}, \hat{v}^{(n)}, \hat{p}^{(n)}\right),$$
$$\delta^{(n)} = \left(\delta_u^{(n)}, \delta_v^{(n)}, \delta_p^{(n)}\right)$$

and

$$R^{(n)} = \left(R^{u^{(n)}}, R^{v^{(n)}}, R^{p^{(n)}}\right),$$

in which

$$R_i^u = b_i^u - \sum_j A_{j_i}^u u_j,$$
$$R_i^v = b_i^v - \sum_j A_{j_i}^v v_j, \quad j = S,\ W,\ P,\ E,\ N$$

and

$$R_i^p = b_i^p - \sum_{j=W,E} A_{j_i}^C u_j - \sum_{j=S,N} A_{j_i}^C v_j,\ i = 1,\ N$$

are the residuals of the block system at the nth block iteration level. The 0th level indicates the values from the previous outer iteration.

The Algorithm (SICS):

Step 0: Calculate vectors h_i, f_i, k_i, n_i and g_i using Eqs. (7.18)–(7.22)
Step 1: Solve $LY^{(n+1)} = B - AX^{(n)}$ for $Y^{(n+1)}$

a. Given $X^{(n)} = (u^{(n)}, v^{(n)}, p^{(n)})$, calculate $R^{(n)} = (R^{u^{(n)}}, R^{v^{(n)}}, R^{p^{(n)}})$
b. Solve $LY^{(n+1)} = R^{(n)}$ for $Y^{(n+1)}$

i. $\hat{u}_i^{(n+1)} = R_i^u \Big/ A_{P_i}^u,\ i = 1,\ N$
ii. $\hat{v}_i^{(n+1)} = R_i^v / A_{P_i}^v,\ i = 1, N$
iii. $\hat{p}_i^{(n+1)} = R_i^p - \left[A_{W_i}^C \hat{u}_{i-1} + A_{E_i}^C \hat{u}_i + A_{S_i}^C \hat{v}_{i-m} + A_{N_i}^C \hat{v}_i\right], \qquad i = 1, N$

Step 2: Solve $U\delta^{(n+1)} = Y^{(n+1)}$ for $\delta^{(n+1)}$

a. Let $\delta_{p_N}^{(n+1)} = 0$

b. $\delta_{p_i}^{(n+1)} = \hat{p}_i^{(n+1)}/g_i,\ i = N-1,\ 1,\ -1$

c. $\delta_{v_i}^{(n+1)} = \hat{v}_i^{(n+1)} - \left[k_i\delta_{p_i}^{(n+1)} + n_i\delta_{p_{i+m}}^{(n+1)}\right],\ i = 1,\ N$

d. $\delta_{u_i}^{(n+1)} = \hat{u}_i^{(n+1)} - \left[h_i\delta_{p_i}^{(n+1)} + f_i\delta_{p_{i+1}}^{(n+1)}\right],\ i = 1,\ N$

Step 3: Update X by $X^{(n+1)} = X^{(n)} + \alpha\delta^{(n+1)}$

a. $u_i^{(n+1)} = u_i^{(n)} + \alpha\delta_{u_i}^{(n+1)},\quad i = 1,\ N$

b. $v_i^{(n+1)} = v_i^{(n)} + \alpha\delta_{v_i}^{(n+1)},\quad i = 1,\ N$

c. $p_i^{(n+1)} = p_i^{(n)} + \alpha\delta_{p_i}^{(n+1)},\quad i = 1,\ N$

7.8.2 Properties and Benefits of SICS

There are a couple of unique properties of SICS, which deserve much attention.

Firstly, by a closer examination of the solution process, the reader can conclude that no extra storage is necessary even for the coefficients of the vectors *h, f, k, n* and *g*. Although they may be computed once and for all for a complete block solution and stored, they may also be calculated wherever the need arises, if such storage is not tolerable.

Secondly, in this procedure, only 8 *N* multiplicative operations are performed for coefficient calculation and 11 *N* operations for each block inner iteration, excluding the residual calculations. The complexity of the procedure is much favorable compared with the complexity of the former procedures.

Thirdly, although the process is a block-wise or a field-wise one, its implementation can be accomplished in a point-wise fashion as well. This is an important feature which will aid in its efficient application to complicated grid arrangements and unstructured grids.

Point-wise application can be accomplished by performing the block solution by calculating R_1^u, R_1^v and R_1^p followed by updating u_1, v_1, p_1, then by calculating R_2^u, R_2^v and R_2^p followed by updating u_2, v_2, p_2, etc. The algorithm developed in the previous subsection may be modified to invoke the point-wise application as follows:

Algorithm (SICS): Point-wise Application

Step 1: *for i=1,N*

Calculate by R_i^u , R_i^v and R_i^p

Solve $\hat{u}_i^{(n+1)} = R_i^u / A_{P_i}^u$

$$\hat{v}_i^{(n+1)} = R_i^v / A_{P_i}^v$$

$$\hat{p}_i^{(n+1)} = R_i^p - \left[A_{W_i}^C \hat{u}_{i-1} + A_{E_i}^C \hat{u}_i + A_{S_i}^C \hat{v}_{i-m} + A_{N_i}^C \hat{v}_i \right]$$

Calculate vectors coefficients h_i, f_i, k_i, n_i and g_i using Eqs. (7.18)-(7.22)

Step 2: Solve $\delta_{p_i}^{(n+1)} = \hat{p}_i^{(n+1)} / g_i$

$$\delta_{v_i}^{(n+1)} = \hat{v}_i^{(n+1)} - \left[k_i \delta_{pi}^{(n+1)} + n_i \delta_{pi+m}^{(n+1)} \right]$$

$$\delta_{u_i}^{(n+1)} = \hat{u}_i^{(n+1)} - \left[h_i \delta_{pi}^{(n+1)} + f_i \delta_{pi+1}^{(n+1)} \right]$$

Step 3: Update $u_i^{(n+1)} = u_i^{(n)} + \alpha \delta_{u_i}^{(n+1)}$

$$v_i^{(n+1)} = v_i^{(n)} + \alpha \delta_{v_i}^{(n+1)}$$

$$p_i^{(n+1)} = p_i^{(n)} + \alpha \delta_{p_i}^{(n+1)}$$

end i

$t = p_N^{(n+1)}$

for i=1, N

$$p_i^{(n+1)} = p_i^{(n+1)} - t$$

end i

The above properties renders the SICS procedure much distinguishable from the previous procedures and especially from the classical segregated type procedures.

Chapter 8
Applications and Testing

In order to apply any of the procedures described to a fluid flow problem, it is necessary to develop a suitable computer program for this purpose. After the initial stage of writing such a program, the code must be tested on some benchmark problems in order to assure its correctness. A benchmark problem is a problem for which solutions were obtained and their correctness established by some previous research. In addition, a benchmark test problem must possess certain characteristics such as equal or differing mesh sizes, various kinds of boundary conditions and varying physical properties. If the code is successful in obtaining correct solutions for these problems, it is most likely that it will work perfectly for other problems as well.

For this purpose, two benchmark problems widely used in the literature are employed in this book. To assist the reader in the application and testing process, some results and comparisons will be given in detail, accompanied by some useful graphics and tables.

8.1 Benchmark Fluid Flow Problems

Problem 1: The Square Tank problem

We consider a square tank full of water. Water enters the tank at one corner, is deflected by the opposing wall and flows out of the diagonally opposite corner. One large circulation zone is developed within the tank. It is experienced that due to this circulation zone, enormous pressure changes occur in the early stages of the iterative solution. This may cause divergence if proper relaxation is not used. Therefore, this problem is a superb test problem for the durability and toughness for any numerical procedure.

© Springer International Publishing Switzerland 2016

Z. Mazhar, *Fully Implicit, Coupled Procedures in Computational Fluid Dynamics*,
Fluid Mechanics and Its Applications 115, DOI 10.1007/978-3-319-29895-5_8

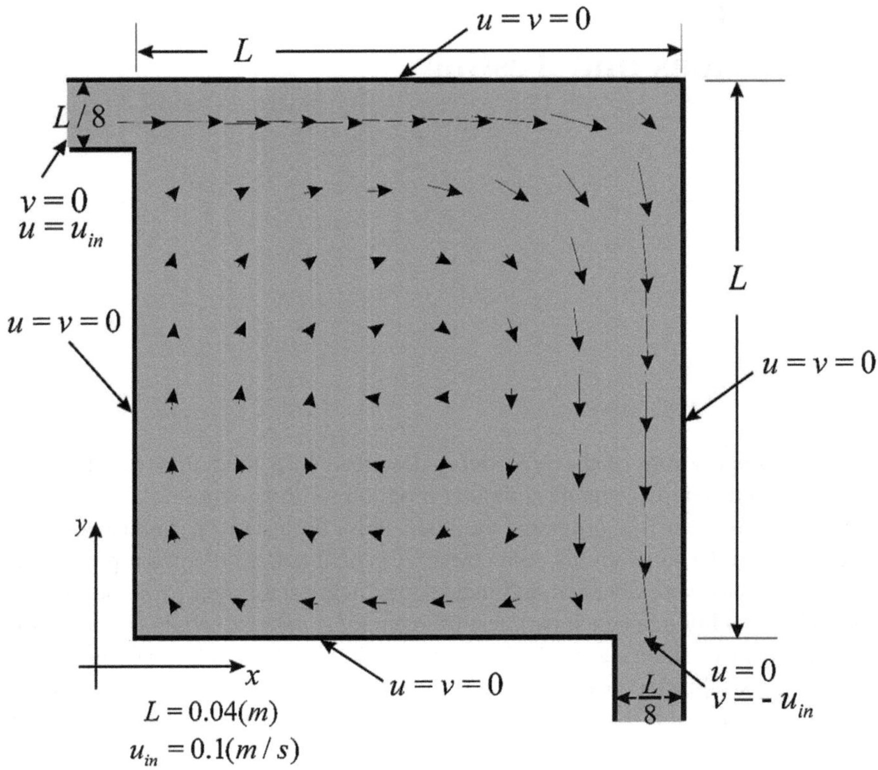

Fig. 8.1 General flow pattern for the Square Tank Problem

Figure 8.1 shows the general flow pattern and the boundary conditions imposed. The physical constants here are those for water; $\rho = 998.2\ \mathrm{kg/m^3}$ and $K_x = K_y = 0.001001$ kg/m−sec. To the interest of an engineer, we note that the Peclet number for this problem is $Pe = \rho u_{in} L/K = 3988.8$.

As shown, only one grid point is located at the center of the inlet and outlet openings and a 10 × 10 interior grid is used. This means that there are 100 interior control volumes or as much pressure points. If the number of grid points in each direction is increased, the number of grid points at the two openings should also be adjusted. For example, if a 20 × 20 grid is utilized, then one more grid point at each opening must be used.

For this problem, equal spacing is used in both directions.

It is much advisable for the reader to start testing the computer code with this problem and with a 10 × 10 grid arrangement. The grid sizes for this case are $\Delta x = \Delta y = 0.004$ m, that is to say $x_i = 0.002 + i\Delta x, i = 0, 9$, $y_i = 0.002 + i\Delta y, i = 0, 9$.

The exact solution to this problem was obtained by driving the solution to a very tight convergence. Numerical results for u, v and p are shown in Tables 8.1, 8.2 and 8.3, for the case of a 10 × 10 grid. In these tables, the indices on the first column and the last row show the pressure point numbers in the y and x directions, respectively.

Table 8.1 Exact *u*-velocity values for the 10 × 10 Square Tank problem

10	0.0998	0.0999	0.0999	0.0993	0.0975	0.0930	0.0837	0.0672	0.0404	0.0000
9	0.0084	0.0148	0.0198	0.0244	0.0298	0.0352	0.0369	0.0321	0.0194	0.0000
8	0.0054	0.0103	0.0136	0.0149	0.0144	0.0143	0.0159	0.0147	0.0090	0.0000
7	0.0041	0.0076	0.0100	0.0110	0.0105	0.0084	0.0074	0.0066	0.0040	0.0000
6	0.0027	0.0049	0.0063	0.0069	0.0064	0.0048	0.0034	0.0026	0.0014	0.0000
5	0.0011	0.0018	0.0022	0.0024	0.0020	0.0016	0.0003	−0.0001	−0.0001	0.0000
4	−0.0008	−0.0019	−0.0027	−0.0031	−0.0028	−0.0028	−0.0033	−0.0024	−0.0009	0.0000
3	-0.0033	−0.0064	−0.0086	−0.0097	−0.0098	−0.0092	−0.0078	−0.0044	−0.0006	0.0000
2	−0.0066	−0.0121	−0.0160	−0.0183	−0.0188	−0.0173	−0.0136	−0.0059	0.0032	0.0000
1	−0.0109	−0.0188	0.0244	−0.0279	−0.0293	−0.0280	−0.0229	−0.0104	0.0243	0.0000
	1	2	3	4	5	6	7	8	9	10

Table 8.2 Exact v-velocity values for the 10 × 10 Square Tank problem

10	0.0000	0.0000	0.0000	0.0000	0.0000	0.0000	0.0000	0.0000	0.0000	0.0000
9	−0.0002	0.0000	0.0000	−0.0005	−0.0018	−0.0045	−0.0093	−0.0165	−0.0267	−0.0404
8	0.0083	0.0064	0.0050	0.0041	0.0035	0.0009	−0.0076	−0.0214	−0.0394	−0.0598
7	0.0137	0.0113	0.0082	0.0055	0.0030	0.0007	−0.0060	−0.0225	−0.0451	−0.0688
6	0.0178	0.0148	0.0106	0.0065	0.0025	−0.0013	−0.0070	−0.0233	−0.0478	−0.0727
5	0.0205	0.0170	0.0121	0.0071	0.0020	−0.0029	−0.0084	−0.0240	−0.0491	−0.0741
4	0.0216	0.0176	0.0125	0.0073	0.0016	−0.0034	−0.0097	−0.0244	−0.0491	−0.0740
3	0.0208	0.0165	0.0117	0.0069	0.0019	−0.0034	−0.0102	−0.0235	−0.0476	−0.0731
2	0.0174	0.0135	0.0095	0.0058	0.0019	−0.0027	−0.0089	−0.0202	−0.0438	−0.0725
1	0.0109	0.0080	0.0056	0.0035	0.0014	−0.0013	−0.0051	−0.0125	−0.0347	−0.0757
	1	2	3	4	5	6	7	8	9	10

Table 8.3 Exact pressure values for Square the 10 × 10 Tank problem

10	−3.7276	−3.7824	−3.8571	−3.9157	−3.9150	−3.7837	−3.4000	−2.6268	−1.4151	0.0000
9	−3.7617	−3.7932	−3.8722	−3.9545	−4.0023	−3.9653	−3.7447	−3.2027	−2.2767	−1.2452
8	−3.8172	−3.8486	−3.9122	−3.9755	−4.0119	−4.0120	−3.9400	−3.6594	−3.1158	−2.5335
7	−3.8748	−3.9060	−3.9616	−4.0140	−4.0424	−4.0398	−4.0051	−3.8767	−3.5986	−3.2980
6	−3.9174	−3.9508	−4.0042	−4.0520	−4.0765	−4.0729	−4.0490	−3.9818	−3.8371	−3.6801
5	−3.9304	−3.9689	−4.0255	−4.0749	−4.0997	−4.0968	−4.0777	−4.0269	−3.9281	−3.8362
4	−3.9014	−3.9497	−4.0148	−4.0704	−4.0986	−4.0946	−4.0677	−4.0047	−3.9159	−3.8670
3	−3.8185	−3.8835	−3.9646	−4.0315	−4.0649	−4.0540	−3.9991	−3.9024	−3.8219	−3.8390
2	−3.6970	−3.7884	−3.8891	−3.9661	−3.9988	−3.9708	−3.8654	−3.7074	−3.6478	−3.8687
1	−3.6075	−3.7201	−3.8313	−3.9104	−3.9349	−3.8815	−3.7040	−3.4023	−3.2914	−4.7066
	1	2	3	4	5	6	7	8	9	10

For the purposes of demonstration and reference, three different grid sizes will be considered for this problem; 10 × 10, 20 × 20 and 40 × 40.

Problem 2: The Shelterbelt problem

In this problem, a dense shelter is placed vertically at the bottom of a flat surface, opposing an approaching velocity field. When the flow encounters the shelterbelt, it must pass over it. High pressure differences are encountered in the vicinity of the top of the shelterbelt. The flow separates at the tip of the shelterbelt and forms a swirling separation bubble between the top of the shelterbelt and a reattachment point, expected at a couple of shelterbelt heights downwind from the shelterbelt. This bubble is characterized by a low velocity reverse flow near the ground. After the reattachment point, the flow readjusts itself towards a profile similar to that at the upstream of the shelterbelt.

The geometry of the problem and the associated boundary conditions are shown in Fig. 8.2.

Assuming that the associated fluid is air, a density of $\rho = 1.18\ \mathrm{kg/m^3}$ is used. At the inflow, an initial boundary condition for u is used as

$$u(y) = \frac{u_*}{\kappa}\ln(y/y_0) \tag{8.1}$$

where $u_* = 0.31\ \mathrm{m/sec}$ is the friction velocity, $\kappa = 0.4$ is the von Karman's constant, $y_0 = 0.007$ m is the roughness parameter and y is the height above the ground. Initially, the u velocity is set as uniform across horizontal lines, whereas the v velocity and pressure are set to zero everywhere.

The choice of the diffusion coefficients K_x and K_y is important because they play a vital role in the flow development. If the flow is turbulent, for example, they will include the effects described by means of a turbulence model through some extra

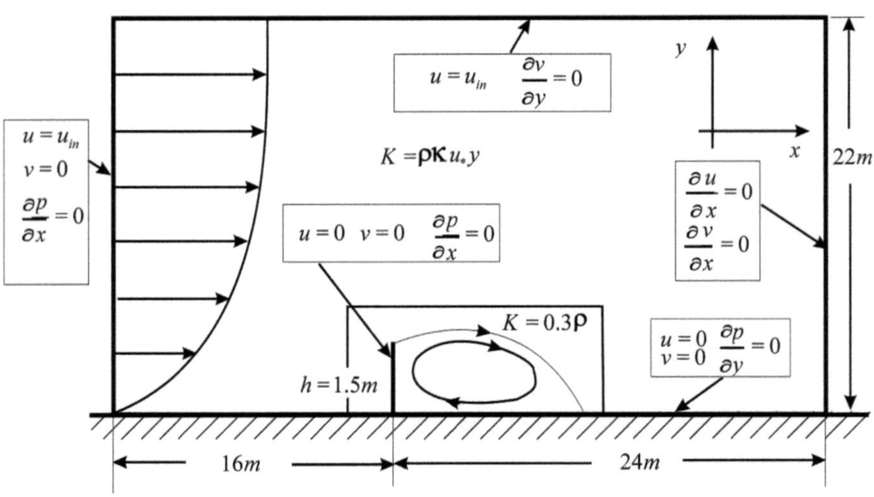

Fig. 8.2 Shelterbelt geometry and boundary conditions

governing partial differential equations. The use of a turbulence model to describe these coefficients does not hinder the way in which the velocity-pressure coupling problem is handled, and therefore is left beyond the scope set up in this book. Instead, we utilize some *semi-constant* diffusion coefficients as described in Fig. 8.2. These were found to yield a *physically reasonable* flow field.

For this problem, significant grid refinements need to be done in the neighborhood of the shelterbelt and near the ground. The grid ratios $(\Delta x/\Delta y)$ for the pressure control volumes may change from 0.1 to 10.

A sample of the grid arrangement for this problem is given in Fig. 8.3.

The shape of the resulting flow field is shown in Fig. 8.4.

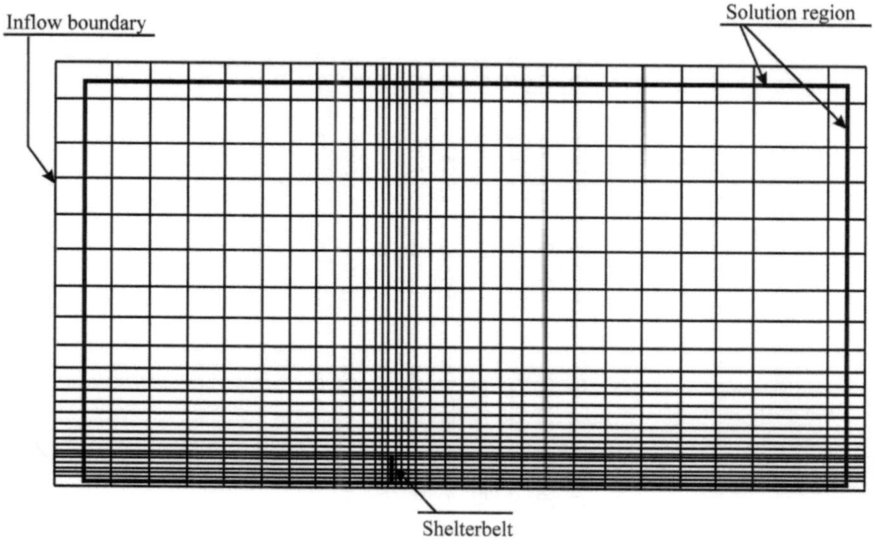

Fig. 8.3 Sample grid arrangement for the Shelterbelt Problem

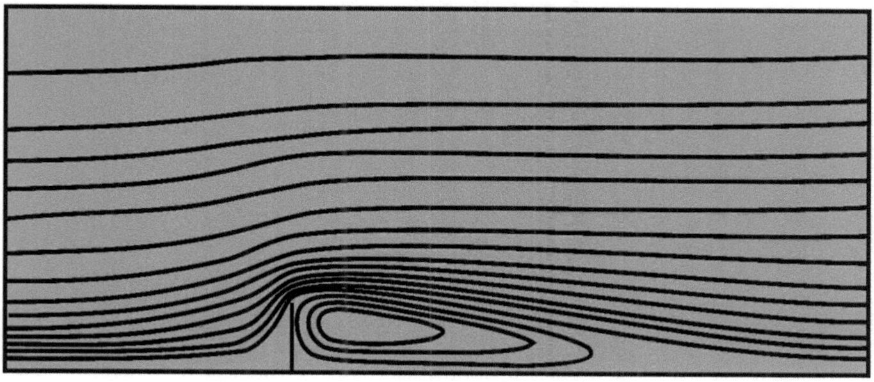

Fig. 8.4 General flow pattern for the Shelterbelt Problem

Table 8.4 Exact u velocity values for 12×9 Shelterbelt problem

9	6.1671	6.1671	6.1671	6.1671	6.1671	6.1671	6.1671	6.1671	6.1671	6.1671	6.1671	6.1671
8	5.9835	5.9845	6.0081	6.0328	6.0578	6.0805	6.1053	6.1340	6.1612	6.1830	6.1943	6.1943
7	5.7808	5.8142	5.8628	5.9042	5.9369	5.9603	5.9865	6.0181	6.0458	6.0698	6.0850	6.0850
6	5.5731	5.6380	5.7213	5.7897	5.8392	5.8711	5.9019	5.9287	5.9433	5.9532	5.9618	5.9618
5	5.2787	5.3745	5.4980	5.6136	5.6985	5.7414	5.7662	5.7574	5.7244	5.6883	5.6692	5.6692
4	4.9917	5.0718	5.1588	5.3026	5.4727	5.4890	5.4258	5.2786	5.1247	5.0115	4.9470	4.9470
3	4.4673	4.3754	4.1601	4.0688	4.5347	4.1533	3.7048	3.2668	3.0031	2.8930	2.8594	2.8594
2	3.6803	3.1223	2.4233	1.5890	0.0000	0.3870	0.7658	1.0909	1.3393	1.4611	1.5039	1.5039
1	2.6216	1.9257	1.3227	0.7137	0.0000	−0.2470	−0.1162	0.2933	0.5967	0.7331	0.7774	0.7774
	1	2	3	4	5	6	7	8	9	10	11	12

Table 8.5 Exact v velocity values for the 12 × 9 Shelterbelt problem

9	0.0479	0.0906	0.0863	0.0762	0.0579	0.0402	0.0397	0.0411	0.0326	0.0189	0.0074	0.0000
8	0.0479	0.0906	0.0863	0.0762	0.0579	0.0402	0.0397	0.0411	0.0326	0.0189	0.0074	0.0000
7	0.0372	0.0917	0.1134	0.1148	0.1093	0.1029	0.0948	0.0813	0.0617	0.0376	0.0165	0.0000
6	0.0455	0.1209	0.1619	0.1717	0.1678	0.1595	0.1458	0.1200	0.0875	0.0556	0.0271	0.0000
5	0.0641	0.1696	0.2333	0.2521	0.2441	0.2256	0.1970	0.1482	0.0993	0.0620	0.0323	0.0000
4	0.0920	0.2295	0.3215	0.3654	0.3529	0.2996	0.2314	0.1405	0.0773	0.0426	0.0228	0.0000
3	0.1279	0.2696	0.3713	0.4782	0.5273	0.3221	0.1613	0.0372	−0.0048	−0.0061	−0.0030	0.0000
2	0.1597	0.2351	0.2790	0.4215	0.8857	−0.0724	−0.2123	−0.1933	−0.1104	−0.0416	−0.0131	0.0000
1	0.1288	0.1305	0.1292	0.1791	0.2745	0.1278	−0.0545	−0.1078	−0.0607	−0.0220	−0.0067	0.0000
	1	2	3	4	5	6	7	8	9	10	11	12

Table 8.6 Exact pressure values for the 12 × 9 Shelterbelt problem

9	0.7716	1.2920	1.7112	1.6277	1.6414	1.5616	0.9645	0.7462	0.6266	0.4461	0.1858	0.0000
8	1.6231	1.6752	1.6540	1.4791	1.2966	1.1123	0.9458	0.7588	0.5413	0.3298	0.0982	−0.0601
7	2.2492	2.1357	1.8546	1.4885	1.1864	0.9487	0.7788	0.5847	0.3513	0.1421	−0.0558	−0.1928
6	2.9906	2.7537	2.2276	1.5974	1.0968	0.7336	0.5006	0.2632	0.0453	−0.0986	−0.2573	−0.4072
5	3.9760	3.5806	2.7847	1.7929	0.9127	0.2764	−0.0542	−0.3222	−0.4003	−0.3684	−0.4580	−0.6538
4	5.3053	4.6104	3.5751	2.2693	0.5894	−0.9369	−1.2907	−1.3582	−0.9385	−0.6013	−0.5797	−0.8124
3	6.9538	5.6145	4.4217	3.4564	0.8159	−5.0728	−3.8245	−2.5248	−1.2176	−0.6082	−0.5669	−0.7970
2	8.7878	6.1031	4.7320	4.5947	4.8540	−6.5730	−4.9872	−2.5091	−0.9776	−0.4361	−0.4989	−0.7596
1	10.0644	6.1244	4.7384	4.8196	5.4598	−6.0473	−5.0715	−2.5528	−0.9304	−0.3895	−0.4791	−0.7496
	1	2	3	4	5	6	7	8	9	10	11	12

This problem and the grid arrangement used, is typical of many fluid flow situations such as flow past various structures, flow over backward facing steps and flow through parallel, symmetrically placed shelters in tunnels. It includes fully developed areas of flow in addition to a recirculating flow zone. In this regard, it is a much suitable test problem which can easily be extended to more complicated flow situations and flow geometries.

Three grid sizes of 12 × 9, 26 × 19 and 39 × 30 are employed for this problem. The exact solutions were obtained by driving the solution to a tight convergence. For reference purposes, numerical results for the resulting velocity and pressure fields with the 12 × 9 grid are shown in Tables 8.4, 8.5 and 8.6.

For the 12 × 9 grid, the x and y coordinates of the grid points utilized are

$$x: (-14.0, -10.0, -6.0, -3.0, -0.9, 0.9, 2.0, 4.5, 7.7, 12.0, 17.0, 22.0)$$

and

$$y: (0.5, 1.0, 2.0, 4.0, 6.0, 9.0, 12.0, 16.0, 20.0)$$

8.2 Testing Criteria

In order to be able to obtain an idea of the performances and of the relative solution economies of the procedures, a suitable testing criterion must be used. In this section, this criterion is set up, followed by the presentation of the preparatory findings to be used in an overall comparison.

There are some considerations that must be taken into account while testing the performance of a procedure.

The cost of the solution of a problem depends mainly on the amount of time required for the complete solution. The total time, on the other hand, depends on various considerations.

First of all, the code writer must be a professional one. A perfect procedure may be ruined by an amateur or by an unexperienced code writer. The sequence of the coding must be designed carefully in order to provide simplicity and to avoid any unnecessary repetitions or difficult-to-follow complex structures. Professional computer scientists for sure can distinguish between statements that spend much time and those which spend less time in execution. Only statements which are easy to follow and consume the least amount of computer time must be preferred. Therefore, the amount of total computer time spent for the complete solution to a problem cannot be considered as the sole criterion for determining the total cost.

Secondly, the amount of time cannot be determined solely by the number of overall iterations performed. The time spent in each block solution varies for each procedure, since the mechanism in each is different. Therefore, the number of block solutions, too cannot be used solely as a criterion for comparison. It can perhaps be used for an analysis of the overall convergence behavior of the procedure.

Perhaps the most decisive factor in this regard is the total number of multiplicative operations performed in the inner iterations of the block solutions. But this in turn, depends on the relaxation parameters used. It is the experience of the Author that, widely varying total costs may result with different relaxation parameters. Therefore, relative comparisons between the procedures must be accomplished under optimum conditions for each individual procedure.

The procedures must first be tested to determine their performance for various values of the relaxation factor E to determine how strong the nonlinearities are taken care of. If the nonlinearities are not treated well, then the required number of overall iterations increase. This, in turn, increases the total number of operations. This process is also necessary to determine the range of this parameter in which convergence or divergence may occur. It should also be noted that the larger the range of E values and the larger the optimal E value giving fastest convergence are measures for the durability and stability of a procedure.

The amount of inner iterations performed to reach a convergence level within the block solution process is mainly affected by the block relaxation parameter α. Once an optimum value for the E parameter is fixed, tests must be conducted to determine the optimum block relaxation parameter α as well. Here again, a procedure can be regarded as more robust if the range of α values leading to convergence is wider and the optimum α value is close to 1. We note that an α value of 1 means that no relaxation is applied.

The number of inner iterations performed in the intermediate block solutions further depends on the level of convergence to be performed in each level. A low level of convergence may result in a slow overall convergence, while a high level will be costly. In fact, it is not necessary to drive these iterations to full convergence since the overall changes would still continue due to the nonlinearities. Therefore, unnecessarily high number of iterations must be avoided.

One remedy of this problem is to suitably fix the number of maximum and minimum inner iterations in the intermediate block solutions. Noting that wild oscillations in the velocity and pressure differences usually occur in the early stages of flow development, this may suitably allow more number of iterations to be performed in these stages. In the meantime it may prevent unnecessary iterations at the later stages.

Another way may be to fix a certain convergence level, suitable for the mass conservation or for the pressure differences by using the respective residual errors. Iterations may be continued until the residuals are dropped by a certain percentage of their original value at the start of the block solution process.

Due to the fact that the changes in the velocity and pressure fields are high in the early stages and low in later stages of the flow development, each one of these criteria has its advantages and disadvantages. In fact, there is no 'best criterion' for this purpose.

In the tests conducted for the purposes of demonstration, a mixture of the above criteria is used. That is to say, while fixing the minimum and maximum number of iterations, pressure differences were also monitored. If the pressure differences have dropped to a prescribed minimal value, the iterations were terminated.

The iterations were also stopped in a case that the minimal value has not been reached, but the maximum number of iterations has been performed.

With the above considerations, to begin with, a series of tests were performed for all problems and for all procedures to determine the range of the convergent E and α values. The range of convergent parameters thus found are shown in Table 8.7.

Following this, finer tests were conducted to determine the optimal relaxation parameters which give the fastest convergence. The total number of operations performed to reach to this level was monitored. The optimal set of the (E, α) couples that lead to the required convergence level with a minimum number of operations were then recorded.

The optimal (E, α) couples found in this way are shown in Table 8.8.

Table 8.7 Range of convergent relaxation parameters

	Square Tank						Shelterbelt						
	10x10		20x20		40x40		12x9		26x19		39x30		
	E	α	E	α	E	α	E	α	E	α	E	α	
BIP	12	0.82	12	0.75	16	0.70	∞	0.92	∞	0.88	500	0.72	max
	1	0.10	1	0.10	1	0.10	1	0.10	1	0.10	1	0.10	min
BIPEN	500	1	30	0.92	12	0.88	∞	0.92	∞	0.90	1000	0.88	max
	1	0.10	1	0.10	1	0.10	1	0.10	1	0.10	1	0.10	min
FICS-1	∞	1	∞	1	∞	1	∞	1.00	∞	1.00	∞	1.00	max
	1	0.10	1	0.10	1	0.10	1	0.10	1	0.10	1	0.10	min
FICS-2	∞	1	∞	1	∞	1	∞	1.00	∞	1.00	∞	1.00	max
	1	0.10	1	0.10	1	0.10	1	0.10	1	0.10	1	0.10	min

Table 8.8 Optimal (E, α) values

	Square Tank						Shelterbelt					
	10x10		20x20		40x40		12x9		26x19		39x30	
	E	α	E	α	E	α	E	α	E	α	E	α
BIP	10	0.80	10	0.70	16	0.70	20	0.92	20	0.88	20	0.65
BIPEN	12	0.75	16	0.85	20	0.85	50	0.90	20	0.75	20	0.70
FICS-1	12	1.00	20	1.00	20	1.00	100	1.00	100	1.00	100	0.90
FICS-2	50	1.00	50	1.00	50	1.00	200	1.00	200	1.00	200	1.00

8.3 Performance Analysis and Comparisons

In order to establish the quality of a procedure, the procedure must be examined on three basic properties: *simplicity*, *durability* and *speed*. Therefore, a performance analysis must be conducted to assure these properties. In this section, a discussion on these aspects is first presented, followed by an investigation of the relative solution economies of the procedures.

For simplicity, the formulation of the procedures must be easy to understand. Furthermore, the implementation must be straightforward. As we have seen, although the construction of the formulations were not perhaps so straightforward, the resulting formulas of the procedures presented are extremely easy.

The simplicity of the formulas reflects itself in the implementations as well. The computer codes for FICS-2 and SICS are given in Appendix C. It can be seen from these codes that very few program lines are necessary for the implementation of the procedures. By examining the algorithm and the coding of SICS, it can be concluded that it is by far the easiest one.

Readers who have an experience with the classical segregated procedures may better appreciate the simplicity of the new procedures thus presented.

The preparatory stages, such as coefficient calculation and boundary condition implementation are almost routine in any solution procedure. However, an advantage with the present procedures is that no pressure boundary condition needs to be imposed onto the mass conservation equations in the course of obtaining the pressures implicitly. In the mass conservation equations, only the setting of the boundary conditions for the velocities is required, but this is quite straightforward.

The elimination of the requirement to form and solve Poisson-type equations is another major simplifying property of the present procedures. This, for sure, is a big leap forward and an important addition to the simplicity.

The memory space requirements for the procedures was discussed in Sect. 7.7. It was seen that not much extra space is required while applying the procedures. The Author is assured that this is the utmost that can be achieved in any solution procedure of such a simplicity. The procedure SICS applied in a point-wise way even requires no extra storage for the block solutions.

In a segregated procedure in which a line-by-line or a point-by-point process is utilized, less storage may be required than some of the block solution procedures presented. However, the amount of space saved is often much superseded by the time required for solution.

The durability of a procedure is another important property that a procedure must possess. In fluid flow problems, a procedure is expected to withstand the destructive effects of any (un)expected and wild changes in the flow field. Moreover, the method must be able to cope with varying grid aspect ratios in all directions. Finally, it must pose strong convergence characteristics.

If the changes in the flow field are not anticipated at all other grid points in an efficient way, then inevitably the mass conservation constraint will be violated. A violation of the incompressibility nature of the fluid, surely must be out of the question, since such a violation will lead to divergence. Such violations, no matter how small they are, must be regarded as 'little time bombs, ready to explode at any time'. The presented procedures deal with this problem in the most efficient way. A discussion on how they cope with this issue is given in Sect. 8.4.

Tests were conducted to determine the behavior of the procedures when widely varying grid aspect ratios was involved. The tests showed that varying grid aspect ratios do not change the character of the convergencies in all of the procedures. Since the symmetric form of the primitive form of the equations was retained and all of the effects of the mass conservation and pressure terms were carried fully to the strength vectors, this was actually an expected behavior.

Another factor in the measurement of the durability of a procedure is the range of the relaxation factors which lead to convergence. The wider the range, the more durable the procedure is. A narrow range normally is close to the lower limit of the relaxation parameter. This means that the procedure requires high under-relaxation, otherwise it will diverge. The Author, in fact, describes relaxation as a rude way of numerically interfering with the natural behavior of flow development.

If the range is narrow, then the user will need to follow a careful but often painful course of fine-tuning for the optimum parameters.

In addition, it must be noted that the higher the value of an optimal relaxation parameter, the least the relaxation is applied. A value of the E factor larger than, say, 1000 virtually means that no relaxation is applied, and an α value equal to 1 means no relaxation is performed in the block solution. If α is 1, then the 'differences' obtained at the end of one inner iteration are fully reflected to the velocity and pressure variables. If α is 0.6, for example, then only sixty percent of the changes are allowed to be reflected. This, for sure, will slow down the convergence.

Now is the time to refer back to Table 8.7. A quick examination of this table leads to the following conclusions:

1. From the last two rows of the table, it can be seen that FICS-2 converges virtually with any value of the relaxation parameters. This means that it may converge even without the need of any relaxation.
2. FICS-1 may require a small amount of relaxation in the inner iterations for very large problems.
3. Both FICS-1 and FICS-2 converge for any value of the relaxation parameter E.
4. The first two rows indicate that BIPEN converges for a wider range of relaxation parameters than BIP.
5. The maximum value of the convergent relaxation parameters decreases for large problems in BIP and BIPEN.
6. All procedures converge even when the lowest allowable values of the relaxation parameters are used.

Overall, it can be concluded that all of the procedures are highly durable, but the durability of FICS-2 is strongest.

Lastly, we turn our attention to the speed property for establishing the quality of the procedures. For this, all of the problems were solved by each procedure, using

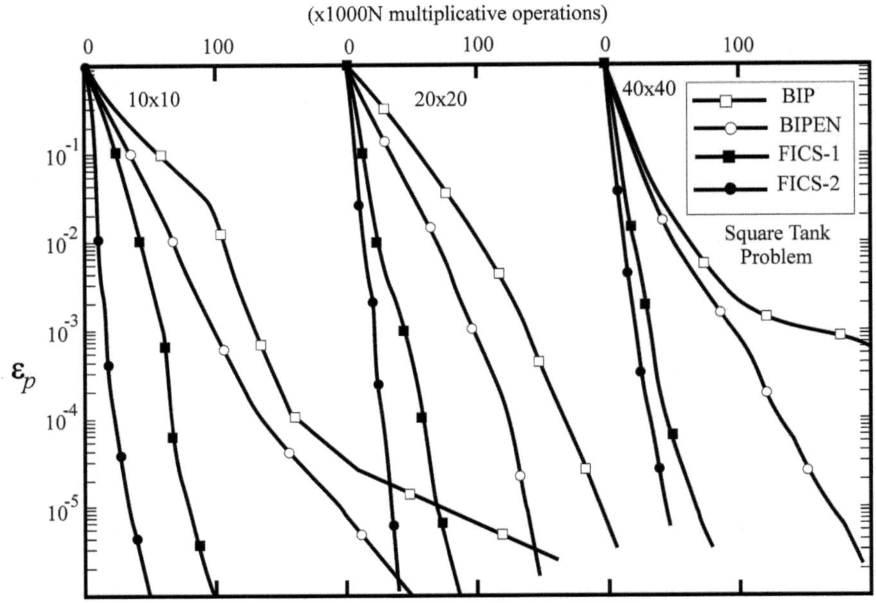

Fig. 8.5 Convergence characteristics versus number of operations performed for the Square Tank problem

the optimum relaxation couples given in Table 8.8. The overall iterations were continued until full convergence was obtained.

Figures 8.5 and 8.6 show the number of operations performed versus the convergence level reached, for the Square tank problem and the Shelterbelt problem, respectively.

The following conclusions can be drawn from these figures:

1. BIP and BIPEN are slower than the other procedures. The speed of these procedures tends to decrease significantly for larger problems.
2. FICS-2 outperforms all of the other procedures, both in speed and in convergence characteristics.

To provide a better insight into the relative performances of the procedures, the total number of multiplicative operations to reach a convergence level of $\varepsilon \leq 10^{-5}$, using the criterion given in Eq. (7.17), was recorded. Table 8.9 gives the relative costs, in which the total number of operations in each column was divided by the smallest number on the same column.

The following conclusions can be drawn from this table:

1. FICS-2 is much faster than the other procedures in all cases.
2. BIP is slower than the other procedures in all cases.
3. FICS-2 exhibits much better performance for larger problems.
4. FICS-2 is up to ten times faster than BIP.

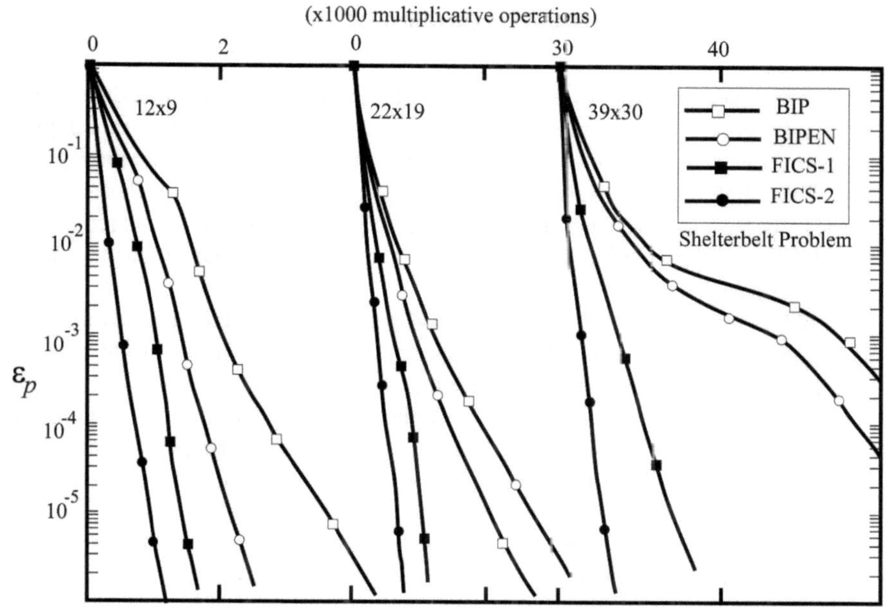

Fig. 8.6 Convergence characteristics versus number of operations performed for the Shelterbelt problem

Table 8.9 Cost comparisons

	Square Tank			Shelterbelt		
	10x10	20x20	40x40	12x9	26x19	39x30
BIP	5.82	5.69	9.63	2.85	4.37	9.91
BIPEN	4.09	3.80	4.40	2.75	3.77	7.92
FICS-1	1.89	1.06	1.85	1.05	1.56	1.88
FICS-2	1	1	1	1	1	1

In conclusion, for the benchmark test problems solved, it can be observed that all of the procedures converge with reasonable cost. The FICS-2 procedure is extremely faster than the other procedures. FICS-2 converges in all cases without needing any relaxation. The most interesting is that, although all of the procedures are adequately durable, FICS-2 surpasses the others by far most. The discussion presented in the following section may give further insight into the superiority of FICS-2.

8.4 A Discussion of the Mechanism of the Procedures

It has been seen in the previous section that all the presented procedures show reasonably good performances in dealing with the velocity-pressure coupling problem. However, the convergence characteristics vary for each solution procedure. The durability and the cost of the solution depend mainly on these characteristics. These characteristics in turn relate to the mechanisms powered by the formulations of the strength vectors used in the matrices L and U and the defect vectors resulting in D.

To obtain a better idea of why these good performances were obtained and why FICS-2 performs much better than the other procedures, it is necessary to examine the mechanisms powered by these procedures.

First of all it is obvious to see that if the effects of the strength vectors are stronger and those of the defect vectors are weaker, then the performance of a procedure will be better. Therefore, it is now suitable to examine the procedures in this regard.

A close examination of the equations of the strength vectors in the four procedures reveals the followings:

1. In all of the procedures, the effects of the main diagonal coefficients of the two momentum equations and the effects of the coefficients of the pressure terms in these equations are equally carried down to the strength vectors h, f, k and n. The term G combines these effects with the effects of the mass conservation coefficients A^C. All of these effects are then transferred down to all of the other strength vectors in a cascaded manner. In this way, the effects of the pressure changes and those of the momentum conservation are harmonized with the constraint imposed by the mass conservation in an efficient way. This is the main reason for the good performances of the procedures. The performance of each procedure now depends on how this process is continued after the formulation of the term g.
2. The procedure BIPEN differs from BIP in the sense that the defect vector q in BIP is completely removed from the defect matrix and transferred as is, as a strength vector, to L as the vector a in BIPEN. The reader may confirm this by noting that the formula for the coefficients of the vector q of BIP is the same as that of vector a of BIPEN. Now in BIPEN, instead of the removed defect vector q, came two other defect vectors e and d just above and below the position of the emptied place of q. The total effect of the i-th coefficients of the vectors e and d is

$$e_i + d_i = -q_i(t_{i-m} + s_{i-m}) \tag{8.2}$$

Therefore the effect of e and d in BIPEN will be less than the effect of q in BIP if

$$|(s + t)_{i-m}| < 1. \tag{8.3}$$

A closer examination of the coefficients resulting in the Square Tank problem, for example, reveals this sum to be $\frac{1}{2}$. This is why the BIPEN procedure is almost twice as fast and much more robust than BIP, as can be seen in Table 8.9.

Now comparing BIPEN and FICS-1, we note first that the strength vectors a and the defect vectors d are the same. In FICS-1, the coefficients of the main diagonal strength vector g in L is fortified by the sum $a_i c_{i-m} + b_i s_{i-1}$. Furthermore, the coefficients of the strength vectors in FICS-1 contain all effects from the coefficients of the mass conservation vectors in a symmetric fashion, a property which BIPEN does not hold. The number of strength vectors in FICS-1 is increased by 1 by adding the vector b and the number of defect vectors is decreased by 1 compared to BIPEN. All these enhancements provide the increased speed and durability of FICS-1.

As can be seen from the expressions of the strength vectors, the effect of all of the coefficients of the primitive block matrix is carried in a fully balanced manner in both of FICS-1 and FICS-2. This provided an enhanced stability to these procedures. This helps much in dealing efficiently with problems involving varying grid size ratios in the discretizations.

Comparing FICS-1 and FICS-2, it can be seen that two additional strength vectors are added in FICS-2, while the number of defect vectors remains the same as in FICS-1. Moreover, in FICS-2, the place of the two defect vectors is shifted by one place. These modifications are the major factors of the superior performance of FICS-2 over FICS-1.

These observations are very important since, with an old saying, 'a couple of birds are being killed with one stone': while a set of equations for the pressure field is formed implicitly and automatically by the decomposition process, the most important effects from the momentum equations and *all* effects from the mass conservation equation and those from the pressure differences are carried efficiently into the strength vectors. More important is the fact that these were achieved without making any unrealistic assumptions about the flow field while forming the decomposition. This asserts the full implicitness of the solution procedures thus formed. It also shows that the procedures based on this decomposition are not based on a point-wise or a line-wise one, but on an actual field-wise one. This means that with these procedures, any single change of the velocity or pressure at any point of the solution region immediately affects not only the values at some neighborhood of this point, but also the values at all of the points of the region.

It is also important to note that at the end of each block solution process, mass and momentum conservations are simultaneously satisfied. In this way the strong coupling between the velocity and pressure fields is fully retained and utilized to encourage faster convergence.

The full implicitness thus realized becomes the main factor for the durability and rapid convergence of the procedures.

8.5 Comparison with the Segregated-Type Procedures

A brief history of the solution methods developed for the treatment of the velocity-pressure coupling problem in the last fifty years, is presented in Appendix A. Much of these classical methods are of the explicit or semi-explicit type. Some kind of implicitness has been tried in the past, but the performance of the proposed methods was not much higher than the former ones. The Author feels that the book would have had something missing if a comparison of these methods with the procedures presented in the book is not being provided. This section was prepared to fill this gap.

It has been demonstrated in the past (see Appendix A) that BIP is more than twice as fast compared to one of the best performing classical procedures of the segregated type, namely SIMPLER. FICS-2 then, in consideration of the conclusions drawn in Sect. 8.3, is faster by a factor of *twenty*. This means some 95 percent reduction in computing time.

In Appendix B, a summary of the formulation of two of the segregated-type methods is presented. There are various drawbacks of these methods which are of utmost importance.

Firstly, and most importantly, the 'segregation' performed in these procedures effectively breaks down the strong coupling between the velocity and pressure fields, which was actually the most crucial property of incompressible fluid flow properties.

Secondly, these methods require the solution of some Poisson-type equation. The solution of these types of equations, however, is known to be one of the most challenging problems in science. The difficulty in the solution of such equations increase as the number of fixed boundary conditions decreases. The problem with pressure solutions is worse since pressure can be fixed at *only one* point within the solution region. The relativeness of the pressure field in incompressible flow problems necessitates that the pressure be let *free* at all boundaries of the solution region. Therefore the matrices resulting for the pressure field in segregated-type solution methods are loosely diagonally dominant. The iterative solution methods for solving systems of equations with matrices carrying this property, therefore, are very slow. These methods slow down dramatically with an increase in the number of grid points used.

Thirdly, the formulation of the Poisson-type of equations in segregated-type methods is complex and sometimes confusing. Moreover, these equations are formed based on some type of assumptions made in the velocity field. Any single assumption to be enforced in a procedure is surely an interference into the natural behavior of fluid flow development. This interference affects the performance of the procedure and necessitates an extra effort to balance the resulting deficiencies.

The segregated procedures lack the ability to quickly distribute the effects of the changes in the velocity and pressure values to the whole of the solution region. The numerical result of this is the violation of the mass and momentum conservation principles, which may cause divergence if strong under-relaxation is not applied.

High relaxation applied presses down the changes, but with the penalty of slowing the convergence. For example, the SIMPLER method does not converge for the 40×40 Square Tank problem if the E factor is chosen larger than 7. Even for the 10×10 problem, it diverges if an E factor more than 15 is used. This indicates a very low performance in durability, in comparison with the stunning performance of FICS-2.

Moreover, in the segregated type procedures, in addition to the E factor, relaxation need to be applied in all of the solution procedures for u, v, p and the pressure correction equations. High relaxation reflects itself in a shrinkage of the range of convergent relaxation parameters. This, in turn, necessitates a painful fine-tuning of the relaxation parameters.

The above difficulties are removed completely with the procedures presented in this book. The strong coupling is by no means interrupted. On the contrary, they are encouraged, so that the changes are reflected smoothly to the whole of the flow field. This also avoids any 'swelling' in the control volumes and enhances the speed of convergence.

In the present procedures, the number of required parameters is reduced to two and the range of parameters leading to convergence is vastly widened. Therefore the optimization phase of these procedures requires much less effort and time than the classical segregated procedures.

In the solution procedures presented, no assumptions whatsoever is made towards either for the pressure or velocity fields. As seen, this is reflected in the overall performance of the procedures.

8.6 Convergence Characteristics and Performances of SICS and SIMPLER: A Relative Comparison

The careful reader should have noticed by now that no mention has yet been made about the performance of the SICS procedure. This was actually intentional. This procedure deserves more attention owing to its unique, additional features of unprecedented simplicity, point-wise applicability and its storage saving capability.

On the other hand, the enthusiastic and experienced readers who have been utilizing the various classical segregated type procedures could not yet find an actual, relative comparison with the block solution procedures, thus presented, which may help in their final decision to switch to the new procedures. The book was mainly designed for introducing the new procedures, but suggestions from various colleagues encouraged the Author to include a special section for this purpose.

Considering the limited space allowed, full comparisons of the procedures with widely varying benchmark problems and extensive criteria may not only be impossible, but also unnecessarily frustrating. Therefore a careful decision had to be made for this purpose.

From the various segregated procedures, the most widely used and typical one is the SIMPLER method. Hence a relative comparison of SIMPLER and SICS procedures will be presented here.

The cost of a segregated procedure mainly depends on the linear equation solver utilized for the solution of the resulting Poisson-type equations. The MSIP solver, also introduced in Appendix B, is perhaps one of the best performing solution methods for such equations. Therefore, in order to hold the scales even, the SIMPLER procedure coupled with MSIP will be used for comparison purposes in what follows.

As discussed in Sect. 8.2, there are various factors affecting the convergence behavior of the solution procedures. The relaxation parameters utilized and the level of convergence allowed in the inner iterations play an important role on the overall convergence behavior of the procedures. Basing the comparison on the number of outer iterations would give an indication on the efficiency of the procedures in treating the nonlinearities in the momentum equations, although the role of the overall iterations and the inner iterations are much different in SICS and SIMPLER. For a total cost comparison however, some more meaningful comparison criteria need to be developed.

For this purpose, the two procedures were extensively tested under all combinations of various parameters, and the cases which lead to least number of multiplicative operations were noted. While doing so, most important observations were noted and reported for the avail of the reader.

The two procedures were also applied to different benchmark problems widely used in the literature. Since the relative performance characteristics in all problems have been found to be similar, only results for the square tank problem with varying grid resolutions will be presented.

The first observation was that SICS converges for any E value in the interval $E \in (1, \infty)$ for all grid resolutions. Although the procedure works with E values less than 1 as well, convergence slows down considerably, leading to a significant increase in the total cost. The optimum E values with SICS are 50, 20 and 18 for the 10×10, 20×20 and 40×40 resolutions, respectively. We note that an E value of 50 accounts for only a 2 % suppression, while for an E value of 20, it accounts for 5 %.

Convergence with SIMPLER could only be achieved for $E \in (1, 15)$, $E \in (1, 10)$ and $E \in (1, 8)$ for the 10×10, 20×20 and 40×40 resolutions, respectively. In SIMPLER with E values closer to these upper limits, results start to flicker with no convergence. For values just above the limits, the process diverges immediately. Interestingly enough, SIMPLER converges faster for $E = 4$ for all problems, which accounts to a 20 % suppression of the dependent variables at all stages.

The optimum α parameters for SIMPLER used in MSIP for the solution of the u and v velocities, pressure correction and pressure were all found to be 0.65. SIMPLER diverges with higher α values and it slows down considerably for lower values.

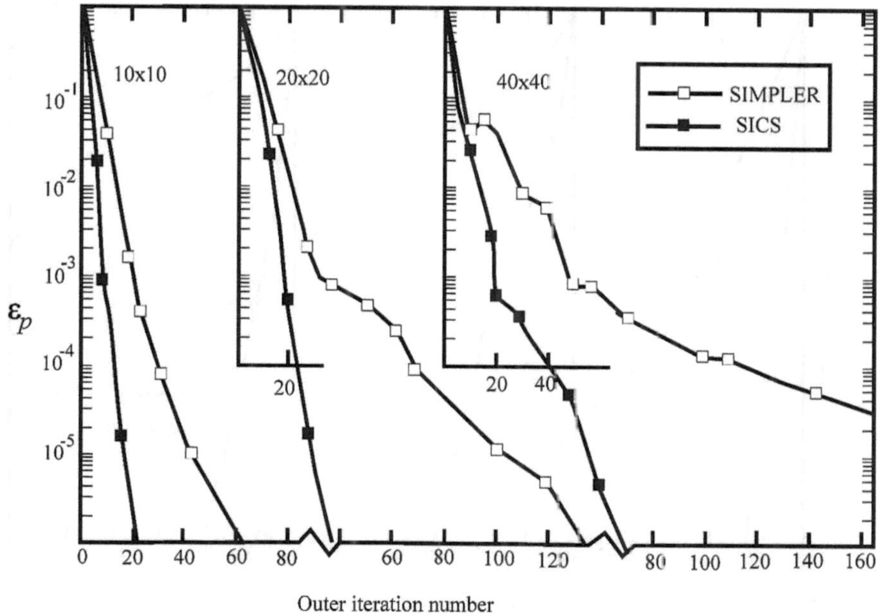

Fig. 8.7 Convergence behavior versus number of outer iterations for SICS and SIMPLER

The optimum α parameter used in the block solutions with SICS is in the range (0.80, 0.90). It was interesting to note that for larger problems SICS would diverge when a α value of 0.9 and a very large E value are used, but convergence with any E value could still be achieved by slightly lowering the α value to 0.85 or 0.8.

Tests were conducted with these optimal parameters and the convergence characteristics of the two procedures were monitored.

In Fig. 8.7 we present the convergence behavior of the two procedures with respect to the number of outer iterations performed.

The number of outer iterations performed to reach an $\varepsilon_p \leq 10^{-5}$ are given in Table 8.10.

The main observation here is that SICS converges in an average of 1.5 m outer iterations and the convergence behavior is almost linear in all cases. SIMPLER however, requires an average of 3.7 m outer iterations, while this number increases

Table 8.10 Total number of outer iterations performed for convergence in SICS and SIMPLER

	10x10	20x20	40x40
SIMPLER	45	111	250
SICS	15	28	59

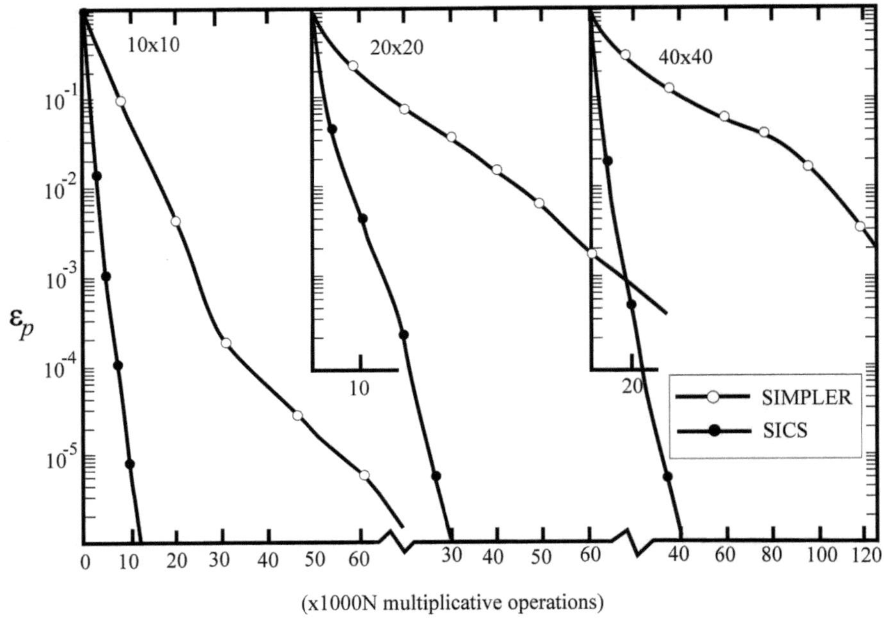

Fig. 8.8 Convergence behavior versus number of operations for SICS and SIMPLER

considerably when m gets larger. This indicates the superiority of SICS over SIMPLER in dealing with the nonlinearities in the momentum equations.

The above observations constitute a clear idea on the relative performances of the procedures on an overall iteration base, but, as we mentioned previously, do not give an idea about the actual relative cost. For the actual cost comparison, we present the total number of multiplicative operations performed to reach a certain convergence level. The results are shown in Fig. 8.8.

In this figure we can appreciate an extremely balanced amount of operations with SICS, a property which is not present with SIMPLER.

The relative cost of each procedure to reach convergence to $\varepsilon_p \leq 10^{-5}$ is shown in Table 8.11, in which the entries were obtained by dividing the number of operations in each column by the one with the least amount.

Table 8.11 Cost comparison for SICS and SIMPLER

	10x10	20x20	40x40
SIMPLER	5.8	7.8	13.7
SICS	1	1	1

This table proves an absolute superiority in computational cost for the SICS procedure. It is observed that the relative cost of SIMPLER increases considerably for larger problems.

Both procedures were tested furthermore with a 100 × 100 grid and it was observed that SICS performed about 20 times faster than SIMPLER. It is evident that for larger problems, more than ninety five percent reduction in computing time may be realized with SICS as compared to SIMPLER.

The poor performance of SIMPLER in distributing the effects of the changes in the velocity and pressure fields, especially when a large number of grid nodes are involved, is clearly attested by Table 8.11.

It is no surprise that SICS converges regardless of what value for the E parameter is used. This is due to the fact that the mass and momentum conservation constraints are satisfied simultaneously at the end of each block solution. This is a demonstration of the robustness of the procedure. We note that this satisfaction is not pursued in the intermediate stages of solution with SIMPLER, but rather expected if and when full convergence is achieved.

As expected, faster convergence rates can be achieved with some optimum values of the relaxation parameters. The optimum E values for SICS are much higher than that of SIMPLER, which is an assurance of the outperforming ability of SICS to cope with abrupt changes from one block solution to another. We note that a larger E value means less under-relaxation, which in turn amounts to less suppression of changes in the flow field.

The high optimum value of 0.8–0.9 for the α relaxation parameter used in the block solution process is another indication of the robustness of SICS. This amounts to a small amount of suppression to deal with the adverse effect of the four defect vectors. When larger problems are considered, inevitably a little more suppression may be required in the block solutions, but this will not be comparable with that required for SIMPLER.

Further tests conducted proved that SICS is also fully insensitive to widely varying grid aspect ratios used in the solution region, a property which SIMPLER does not hold.

Among other advantages of SICS over SIMPLER, such as simplicity and robustness, we may count the advantage that no extra storage is needed in the block solutions.

Chapter 9
Special Cases

In order that a new procedure gain high utilization in research, it is necessary that it holds the ability to be applied to various more general, as well as to some special cases. It should be noted that the book was designed to present the newly developed procedures and to demonstrate their ability to resolve one of the most troublesome problems in computational fluid dynamics, namely the velocity-pressure coupling problem. No aim was really sought in presenting solutions to some specialized physical problems. The space allocated for the book is not adequate for a full pledged investigation or presentation for all possible applications, but the omission of some brief ideas in these various aspects would render it somehow incomplete.

For this purpose, in this chapter, certain topics which were not discussed in the previous chapters will be presented for the sake of completeness and for the reference of the reader. It is hoped that the introductory ideas and guidelines presented on how the adaptations are performed will serve as an aid to researchers, while asserting the applicability of the procedures to further cases.

9.1 Time-Dependent Problems

The governing differential equations whose solutions have been discussed are for steady state problems. If the flow is unsteady or in other words time-dependent, then the governing differential equations (Eqs. (3.1–3.3)) will read

$$\frac{\partial}{\partial x}\left(K_x\frac{\partial u}{\partial x}\right)+\frac{\partial}{\partial y}\left(K_y\frac{\partial u}{\partial y}\right)-\frac{\partial}{\partial x}(\rho uu)-\frac{\partial}{\partial y}(\rho uv)-\frac{\partial p}{\partial x}=\frac{\partial}{\partial t}(\rho u) \tag{9.1}$$

(u momentum equation)

© Springer International Publishing Switzerland 2016
Z. Mazhar, *Fully Implicit, Coupled Procedures in Computational Fluid Dynamics*, Fluid Mechanics and Its Applications 115, DOI 10.1007/978-3-319-29895-5_9

$$\frac{\partial}{\partial x}\left(K_x\frac{\partial v}{\partial x}\right)+\frac{\partial}{\partial y}\left(K_y\frac{\partial v}{\partial y}\right)-\frac{\partial}{\partial x}(\rho uv)-\frac{\partial}{\partial y}(\rho vv)-\frac{\partial p}{\partial y}=\frac{\partial}{\partial t}(\rho v) \tag{9.2}$$

(v momentum equation)

$$\frac{\partial u}{\partial x}+\frac{\partial v}{\partial y}=0 \quad \text{(Mass conservation equation)} \tag{9.3}$$

in which t represents time.

In order that this system govern a true physical problem, in addition to the boundary conditions for the pressure and velocity components, initial profiles for the velocity components must be supplied. Let the initial values be given by

$$u^0 = u(x, y, t = t_0), \quad v^0 = v(x, y, t_0) \tag{9.4}$$

for all (x, y) values in the solution region and for some initial time $t = t_0$.

Noting the relativeness of the pressure field in an incompressible fluid flow and the special treatment of this fact in the solution procedures presented, pressures may be set to zero at the beginning of the block solution. In fact, the fluid is assumed at a stand still at the beginning.

Given these initial values, it is required to find the values of the velocity and pressure variables at a time $t + \Delta t$, where Δt represents a small specified time step. The process may be repeated until a specified time is reached. We denote the values of the velocity and pressure variables at this new time level by $u^{t+\Delta t}$, $v^{t+\Delta t}$ and $p^{t+\Delta t}$.

The discretization of the governing equations is in parallel to the process described in Chap. 4. Discretization of the terms on the left hand side of the momentum equations is the same with one exception: Since now the real time-dependent solutions are sought, relaxation employed by the E factor is not necessary. The solutions, starting from an initial time, are obtained by using a time step Δt and marching in the forward direction to obtain the velocity and pressure fields as time progresses. This can be achieved simply by setting the parameter E to a very large value. If the time step Δt is kept sufficiently small, no sudden differences will be produced in the velocity and pressure fields and the solution process will be stable.

The discretization of the mass conservation equation is also the same. For the discretization of the momentum equations, the terms on the right hand sides of the momentum equations are approximated by

$$\left.\frac{\partial(\rho u)}{\partial t}\right|_P = \rho\left.\frac{\partial u}{\partial t}\right|_P \approx \rho\frac{u_P^{t+\Delta t} - u_P^{t}}{\Delta t} \tag{9.5}$$

and

$$\left.\frac{\partial(\rho v)}{\partial t}\right|_P = \rho\left.\frac{\partial v}{\partial t}\right|_P \approx \rho\frac{\overset{t+\Delta t}{v_P} - \overset{t}{v_P}}{\Delta t} \tag{9.6}$$

To visualize the changes to be made, we rewrite here the discretized forms of the momentum and mass conservation equations for Eqs. (4.32), (4.39) and (4.44) as

$$A_S^u u_S + A_W^u u_W + A_P^u u_P + A_E^u u_E + A_N^u u_N + A_P^{pu} p_P + A_E^{pu} p_E = b_p^u \tag{9.7}$$

$$A_S^v v_S + A_W^v v_W + A_P^v v_P + A_E^v v_E + A_N^v v_N + A_P^{pv} p_P + A_N^{pv} p_N = b_P^v \tag{9.8}$$

$$A_W^{Cu} u_W + A_P^{Cu} u_P + A_S^{Cv} v_S + A_P^{Cv} v_P = b_P^p \tag{9.9}$$

Noting that to obtain these equations, we have multiplied both sides of the discretized equations by $\Delta x\Delta y$. Therefore, we must also multiply the approximations from Eqs. (9.5) and (9.6) by $\Delta x\Delta y$ before substituting them instead of the b_p^u and b_P^v terms in the right hand sides of Eqs. (9.7) and (9.8), respectively. This gives

$$A_S^u u_S + A_W^u u_W + A_P^u u_P + A_E^u u_E + A_N^u u_N + A_P^{pu} p_P + A_E^{pu} p_E = \rho\Delta x\Delta y\frac{\overset{t+\Delta t}{u_P} - \overset{t}{u_P}}{\Delta t} \tag{9.10}$$

$$A_S^v v_S + A_W^v v_W + A_P^v v_P + A_E^v v_E + A_N^v v_N + A_P^{pv} p_P + A_N^{pv} p_N = \rho\Delta x\Delta y\frac{\overset{t+\Delta t}{v_P} - \overset{t}{v_P}}{\Delta t} \tag{9.11}$$

$$A_W^{Cu} u_W + A_P^{Cu} u_P + A_S^{Cv} v_S + A_P^{Cv} v_P = b_P^p \tag{9.12}$$

The velocity and pressure terms on the left hand sides are written for the time $t + \Delta t$, so without any loss in meaning we drop the superscript $t + \Delta t$ from the velocity components on the right hand sides of Eqs. (9.10)–(9.11) and write the set of equations as

$$A_S^u u_S + A_W^u u_W + A_P^u u_P + A_E^u u_E + A_N^u u_N + A_P^{pu} p_P + A_E^{pu} p_E = \rho\Delta x\Delta y\frac{u_P - \overset{t}{u_P}}{\Delta t} \tag{9.10$'$}$$

$$A_S^v v_S + A_W^v v_W + A_P^v v_P + A_E^v v_E + A_N^v v_N + A_P^{pv} p_P + A_N^{pv} p_N = \rho\Delta x\Delta y\frac{v_P - \overset{t}{v_P}}{\Delta t} \tag{9.11$'$}$$

$$A_W^{Cu} u_W + A_P^{Cu} u_P + A_S^{Cv} v_S + A_P^{Cv} v_P = b_P^p \tag{9.12'}$$

Now we collect terms and write these equations as

$$\begin{aligned} & A_S^u u_S + A_W^u u_W + \left[A_P^u - \frac{\rho \Delta x \Delta y}{\Delta t}\right] u_P + A_E^u u_E + A_N^u u_N + A_P^{pu} p_P + A_E^{pu} p_E \\ & = -\frac{\rho \Delta x \Delta y}{\Delta t} u_P^t \end{aligned} \tag{9.13}$$

$$\begin{aligned} & A_S^v v_S + A_W^v v_W + \left[A_P^v - \frac{\rho \Delta x \Delta y}{\Delta t}\right] v_P + A_E^v v_E + A_N^v v_N + A_P^{pv} p_P + A_N^{pv} p_N \\ & = -\frac{\rho \Delta x \Delta y}{\Delta t} v_P^t \end{aligned} \tag{9.14}$$

$$A_W^{Cu} u_W + A_P^{Cu} u_P + A_S^{Cv} v_S + A_P^{Cv} v_P = b_P^p \tag{9.15}$$

The implementation of the above process is performed easily by updating the main diagonal coefficients and the right hand sides of the block system prior to the application of the boundary conditions as follows:

$$\begin{aligned} A_P^u &\leftarrow A_P^u - \frac{\rho \Delta x \Delta y}{\Delta t} \\ b_P^u &\leftarrow b_P^u - \frac{\rho \Delta x \Delta y}{\Delta t} u_P^t \\ A_P^v &\leftarrow A_P^v - \frac{\rho \Delta x \Delta y}{\Delta t} \\ b_P^v &\leftarrow b_P^v - \frac{\rho \Delta x \Delta y}{\Delta t} v_P^t \end{aligned}$$

We remind that the E parameter need to be set to a very large value before the above operations are performed. No other change need to be done in the formulations.

The next step involves the application of the boundary conditions. This step is performed in the same manner as applied in the steady-state case. The assembly process is also the same, giving the coupled block system of equations to be solved in order to obtain the values of the velocity and pressure variables at the time $t + \Delta t$. This process may now be repeated until a required time has been reached.

Time dependency of the problem does not affect the way in which the velocity-pressure problem is handled. Therefore, no change in the block solution scheme is required.

9.2 Stoke's Flow Equations

A special form of the governing differential equations is when the terms containing first order derivatives regarding velocities in the momentum equations are not present. The equations thus formed are called the Stoke's flow equations and read as follows, in which the mass conservation equation remains the same.

$$\frac{\partial}{\partial x}\left(K_x \frac{\partial u}{\partial x}\right) + \frac{\partial}{\partial y}\left(K_y \frac{\partial u}{\partial y}\right) - \frac{\partial p}{\partial x} = 0 \tag{9.16}$$

$$\frac{\partial}{\partial x}\left(K_x \frac{\partial v}{\partial x}\right) + \frac{\partial}{\partial y}\left(K_y \frac{\partial v}{\partial y}\right) - \frac{\partial p}{\partial x} = 0 \tag{9.17}$$

$$\frac{\partial u}{\partial x} + \frac{\partial v}{\partial y} = 0 \tag{9.18}$$

These equations are obtained simply by setting $\rho = 0$ in the momentum equations. Therefore, in such a situation, the discretization equations obtained for the general case will still be valid if they are updated by setting $\rho = 0$. Since now there exist no nonlinearities in the governing equations, the E-factor formulation need not be applied. To cancel the relaxation induced by the E-factor, it is not necessary to make any changes in the formulations. This can be achieved simply by setting the E-factor to a very large value.

However, now, the profile assumptions made in Sect. 4.3 will no longer be valid. This problem is solved simply by setting the coefficients β in the profile assumptions to 1.

With the above three modifications, the general discretization equations apply to Stoke's flow problems as well. The way in which the assembly process is performed and the technique used for solving the block system remain the same.

The velocity and pressure fields are obtained in only one single block solution to a required accuracy.

In summary, it can be concluded that the solution of the Stoke's flow problem is straightforward and can be regarded as a special case.

9.3 Turbulent Flows and Heat Transfer

The mathematical description of fluid flows involving turbulence or heat transfer involves some extra partial differential equations to be introduced, together with the original mass conservation and momentum equations. These extra equations depend on the model selected in order to adequately describe the character of turbulence of the fluid and that of the heat present, and may vary in number. The effects of turbulence and heat transfer are reflected by the terms K_x and K_y in the second order derivative terms of the momentum equations. This difference does not affect the way in which the velocity-pressure coupling problem is handled, therefore the solution procedures already presented remain the same.

In addition to the solution process, the extra equations described by the turbulence and heat transfer need to be solved prior to coefficient refreshment in the overall iterations. The diffusion terms obtained from the solution of these equations are used in the formulation of the descretization of the momentum equations in the normal way.

The solution of the extra equations does not pose an important problem since each of these equations describes a certain single variable. These equations are usually complemented by some strong boundary conditions, which asserts that the involved matrices will be strongly diagonally dominant. Therefore, their solution may be accomplished efficiently by the use of classical linear equation solvers.

9.4 Adaptation to Existing Codes

An engineer who has a working code utilizing any of the classical procedures may tend to hesitate in trying to adapt his code to a completely new approach. This may require some valuable time and perhaps any unsuccessful attempts can cause disappointment. However, this is not the case with the present procedures, since the adaptation process for these procedures is quite fast and easy. To aid the user in this respect, two working subroutines for the block solution procedures FICS-2 and SICS are given in Appendix C. The only preparation required for adaptation is the creation of the additional four vectors for mass conservation and four vectors for pressure differences, which is straightforward. Some care must surely be given to the correct implementation of the boundary conditions in the block system.

While adapting the new procedure, the reader will need to throw away all the codes for the preparatory stages for the u, v, p and p', including boundary condition application for each and the linear solver used to solve the resulting equations. This surely, will result in an immense amount of reduction in the coding, to the surprise of the reader.

9.5 Three-Dimensional Problems

An additional advantage of the presented procedures is that they can be easily adapted to the solution of three-dimensional problems as well. For such problems, a third governing equation for momentum is added and the mass conservation equation contains a third term. The equations are given as

$$\frac{\partial}{\partial x}\left(K_x\frac{\partial u}{\partial x}\right)+\frac{\partial}{\partial y}\left(K_y\frac{\partial u}{\partial y}\right)+\frac{\partial}{\partial z}\left(K_z\frac{\partial u}{\partial z}\right)-\frac{\partial}{\partial x}(\rho uu)-\frac{\partial}{\partial y}(\rho uv)$$
$$-\frac{\partial}{\partial z}(\rho uw)-\frac{\partial p}{\partial x}=\frac{\partial}{\partial t}(\rho u) \tag{9.19}$$
(u momentum equation)

$$\frac{\partial}{\partial x}\left(K_x\frac{\partial v}{\partial x}\right)+\frac{\partial}{\partial y}\left(K_y\frac{\partial v}{\partial y}\right)+\frac{\partial}{\partial z}\left(K_z\frac{\partial v}{\partial z}\right)-\frac{\partial}{\partial x}(\rho vu)-\frac{\partial}{\partial y}(\rho vv)$$
$$-\frac{\partial}{\partial z}(\rho vw)-\frac{\partial p}{\partial y}=\frac{\partial}{\partial t}(\rho v) \tag{9.20}$$
(v momentum equation)

$$\frac{\partial}{\partial x}\left(K_x\frac{\partial w}{\partial x}\right)+\frac{\partial}{\partial y}\left(K_y\frac{\partial w}{\partial y}\right)+\frac{\partial}{\partial z}\left(K_z\frac{\partial w}{\partial z}\right)-\frac{\partial}{\partial x}(\rho wu)-\frac{\partial}{\partial y}(\rho wv)$$
$$-\frac{\partial}{\partial z}(\rho ww)-\frac{\partial p}{\partial z}=\frac{\partial}{\partial t}(\rho w) \tag{9.21}$$
(w momentum equation)

$$\frac{\partial u}{\partial x}+\frac{\partial v}{\partial y}+\frac{\partial w}{\partial z}=0 \quad \text{(Mass conservation equation)} \tag{9.22}$$

The discretization of these equations is similar to the two-dimensional ones. The block matrix system is extended to contain the new additions and will be of the form

$$\begin{bmatrix} A_{11} & 0 & 0 & A_{14} \\ 0 & A_{22} & 0 & A_{24} \\ 0 & 0 & A_{33} & A_{34} \\ A_{41} & A_{42} & A_{43} & 0 \end{bmatrix}\begin{bmatrix} u \\ v \\ w \\ p \end{bmatrix}=\begin{bmatrix} b^u \\ b^v \\ b^w \\ b^p \end{bmatrix}$$

The block submatrices have the following form:

$A_{11} = A_{22} = A_{33} =$ 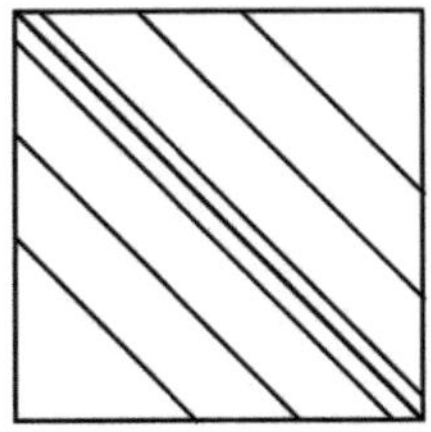 $A_{14} =$ 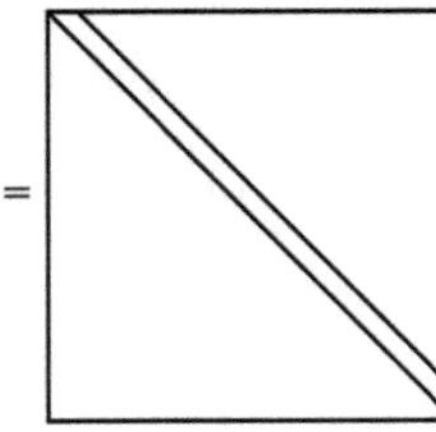

$A_{24} =$ 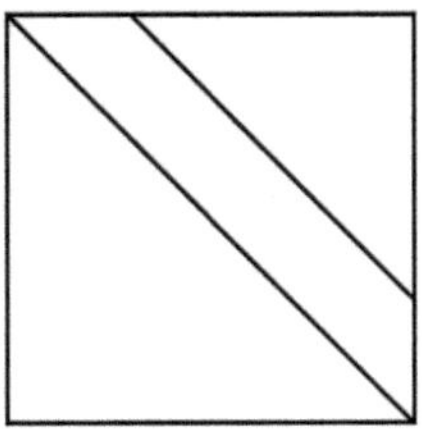 $A_{34} =$ 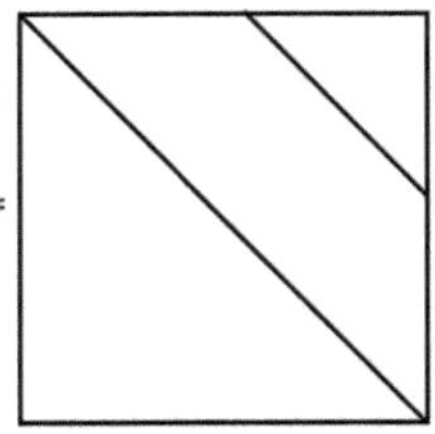

$A_{41} =$ 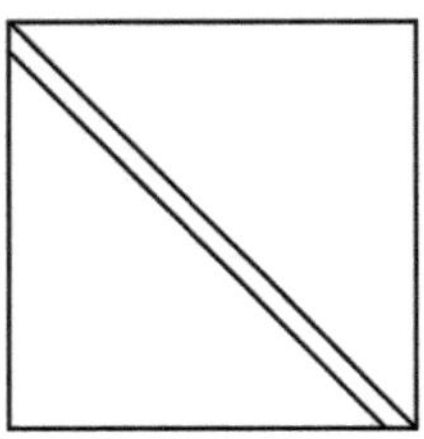 $A_{42} =$ 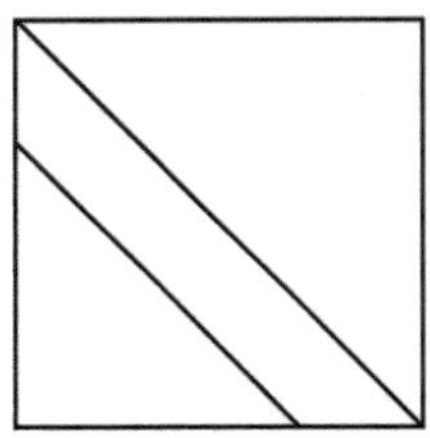 $A_{43} =$ 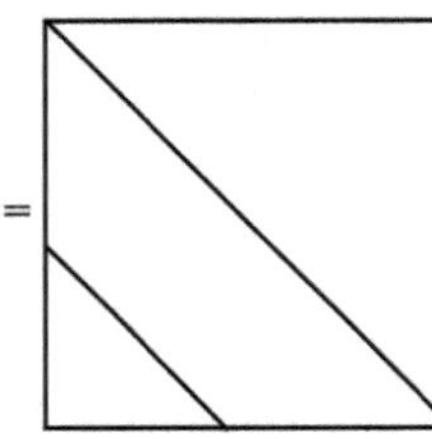

The submatrices L and U are formed similar to what was done in the two-dimensional case. All the off diagonals of the momentum equations are transferred to the corresponding places in the defect matrix. The submatrices corresponding to pressure and mass conservation are retained as they appear in their primitive forms.

What remains is the placement of the rest of the strength vectors in the L_{44} and U_{44} submatrices, which can be done in various ways as in two dimensions. The places of the vectors in the submatrix D_{44} then will be determined according to the placement of the respective strength vectors.

When FICS-2 is used, for example, the places of the strength vectors in the L_{44} and U_{44} submatrices will be as follows:

$L_{44} =$ 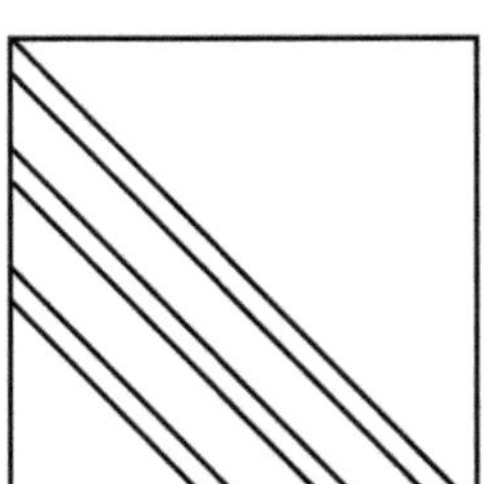 $U_{44} =$ 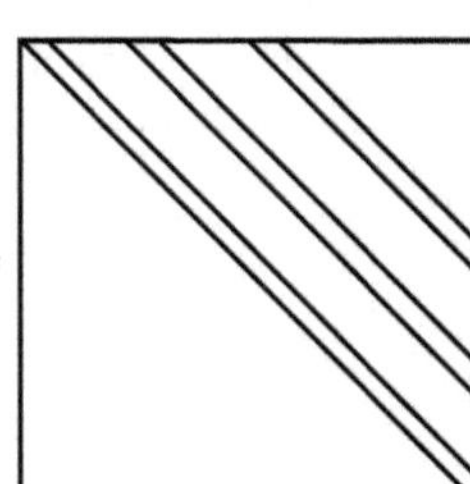

We leave the determination of the places of the defect vectors and the formulas for the coefficients of these vectors for FICS-2 to the interested reader.

To provide an insight to the reader, however, we provide a full presentation of the formulas for the SICS procedure below.

Assume that in addition to the notations used previously for SICS in two dimensions, we introduce $A^w_{P_i}$ as the main diagonal vector coefficient for the w variable. We also use $A^p_{D_i}$ and $A^p_{U_i}$ to denote the main diagonal and the far off diagonal coefficients, respectively, for the pressure differences in the third momentum equation. We further introduce $A^C_{U_i}$ and $A^C_{D_i}$ to denote the main diagonal and the far off diagonal vector coefficients, respectively, for the mass conservation equation. We note that these coefficients contain only the grid sizes in the z direction. In addition, we let the main diagonal and far off diagonal coefficients of the strength vectors in the matrix U referring to the third momentum equation, be denoted by t_i and s_i, respectively. The total number of grid points is $N = m \times n \times l$ in which l is the number of grid points employed in the z direction.

Then it is relatively easy to visualise the formulas for the coefficients of the strength vectors h, f, k, n, t, s and g. They are given as

Formulas for SICS: Three dimensional problems:

$$h_i = A^p_{W_i}/A^u_{P_i}, \quad i = 1, N \tag{9.23}$$

$$f_i = A^p_{E_i}/A^u_{P_i}, \quad i = 1, N-1 \tag{9.24}$$

$$k_i = A^p_{S_i}/A^v_{P_i}, \quad i = 1, N \tag{9.25}$$

$$n_i = A^p_{N_i}/A^v_{P_i}, \quad i = 1, N-m \tag{9.26}$$

$$t_i = A^p_{D_i}/A^w_{P_i} \quad i = 1, N \tag{9.27}$$

$$s_i = A^p_{U_i}/A^w_{P_i}, \quad i = 1, N-m-l \tag{9.28}$$

$$g_i = -\left[A^C_{W_i}A^p_{E_i} + A^C_{E_i}A^p_{W_i} + A^C_{S_i}A^p_{N_i} + A^C_{N_i}A^p_{S_i} + A^C_{D_i}A^p_{U_i} + A^C_{U_i}A^p_{D_i}\right], \quad i = 1, N \tag{9.29}$$

Application of the block solution procedure follows similar steps as in the two dimensional case. If preferred, the application of the block solution process may be accomplished in a point-wise fashion as well. To finish with, we give the block-wise algorithm.

We let

$$X^{(n)} = \left(u^{(n)}, v^{(n)}, w^{(n)}, p^{(n)}\right)$$
$$Y^{(n)} = \left(\hat{u}^{(n)}, \hat{v}^{(n)}, \hat{w}^{(n)}, \hat{p}^{(n)}\right),$$
$$\delta^{(n)} = \left(\delta_u^{(n)}, \delta_v^{(n)}, \delta_w^{(n)}, \delta_p^{(n)}\right)$$
$$\text{and } R^{(n)} = \left(R^{u^{(n)}}, R^{v^{(n)}}, R^{w^{(n)}}, R^{p^{(n)}}\right),$$

in which

$$R_i^u = b_i^u - \sum_j A_{j_i}^u u_j,$$
$$R_i^v = b_i^v - \sum_j A_{j_i}^v v_j$$
$$R_i^w = b_i^w - \sum_j A_{j_i}^w w_j,$$

and

$$R_i^p = b_i^p - \sum_{j=W,E} A_{j_i}^{C^u} u_j - \sum_{j=S,N} A_{j_i}^{C^v} v_j - \sum_{j=D,U} A_{j_i}^{C^w} w_j,$$

for $i = 1, N$ and $j = S, W, P, E, N, D, U$.

The Algorithm (SICS) for three dimensions:

Step 0: Calculate vectors h_i, f_i, k_i, n_i, s_i, t_i and g_i using Eqs. (9.23)–(9.29)
Step 1: Solve $LY^{(n+1)} = B - AX^{(n)}$ for $Y^{(n+1)}$

a. Given $X^{(n)} = \left(u^{(n)}, v^{(n)}, w^{(n)}, p^{(n)}\right)$, calculate $R^{(n)} = \left(R^{u^{(n)}}, R^{v^{(n)}}, R^{w^{(n)}}, R^{p^{(n)}}\right)$
b. Solve $LY^{(n+1)} = R^{(n)}$ for $Y^{(n+1)}$

i. $\hat{u}_i^{(n+1)} = R_i^u / A_{P_i}^u, \quad i = 1, N$
ii. $\hat{v}_i^{(n+1)} = R_i^v / A_{P_i}^v, \quad i = 1, N$
iii. $\hat{w}_i^{(n+1)} = R_i^w / A_{P_i}^w, \quad i = 1, N$
iv. $\hat{p}_i^{(n+1)} = R_i^p - \left[A_{W_i}^C \hat{u}_{i-1} + A_{E_i}^C \hat{u}_i + A_{S_i}^C \hat{v}_{i-m} + A_{N_i}^C \hat{v}_i + A_{D_i}^C \hat{w}_{i-m-l} + A_{U_i}^C \hat{w}_i\right], \quad i = 1, N$

Step 2: Solve $U\delta^{(n+1)} = Y^{(n+1)}$ for $\delta^{(n+1)}$

i. Let $\delta_{p_N}^{(n+1)} = 0$
ii. Solve $\delta_{p_i}^{(n+1)} = \hat{p}_i^{(n+1)} / g_i, \quad i = N - 1, 1, -1$

iii. Solve $\delta_{w_i}^{(n+1)} = \hat{w}_i^{(n+1)} - \left[s_i \delta_{p_i}^{(n+1)} + t_i \delta_{p_{i+m+l}}^{(n+1)}\right], \quad i = 1, N$

iv. Solve $\delta_{v_i}^{(n+1)} = \hat{v}_i^{(n+1)} - \left[k_i \delta_{p_i}^{(n+1)} - n_i \delta_{p_{i+m}}^{(n+1)}\right] \quad i = 1, N$

v. Solve $\delta_{u_i}^{(n+1)} = \hat{u}_i^{(n+1)} - \left[h_i \delta_{p_i}^{(n+1)} + f_i \delta_{p_{i+1}}^{(n+1)}\right], \quad i = 1, N$

Step 3: Update X by $X^{(n+1)} = X^{(n)} + \alpha \delta^{(n+1)}$

a. $u_i^{(n+1)} = u_i^{(n)} + \alpha \delta_{u_i}^{(n+1)}, \quad i = 1, N$

b. $v_i^{(n+1)} = v_i^{(n)} + \alpha \delta_{v_i}^{(n+1)}, \quad i = 1, N$

c. $w_i^{(n+1)} = w_i^{(n)} + \alpha \delta_{w_i}^{(n+1)}, \quad i = 1, N$

d. $p_i^{(n+1)} = p_i^{(n)} + \alpha \delta_{p_i}^{(n+1)}, \quad i = 1, N$

Chapter 10
Concluding Remarks

A new family of solution procedures has been introduced with a fresh perspective for the solution of field problems. Among the members of this family, the two most powerful and distinctive ones are FICS-2 and SICS. The FICS-2 procedure is faster than SICS, but with a slight disadvantage that it requires a couple of extra vectors. Luckily enough, the storage is not more than that required by a classical segregated procedure and can be tolerated.

The demonstrated simplicity, versatility, robustness and speed, combined with the elimination of the laborious tasks of the setting up and solving Poisson-type of equations, renders these procedures as efficient alternatives to the classical segregated procedures. Quick adaptation to existing codes together with their direct applicability to unstructured grids and to three dimensional problems with no extra computing memory requirements are some of the additional benefits.

The correct diagnosis of the pressure-velocity coupling problem was the starting point of the adventure. Not perhaps surprisingly, the treatment was based mainly on the most basic principles of fluid dynamics: fluid flow initiated by pressure differences and continued by obeying the mass conservation principle. In this way, the strong coupling between the velocity and pressure fields was not disrupted, but rather utilized as a beneficial factor. The incomplete decomposition and block solution of the primitive form of the discretized equations were merely efficient mechanisms for this purpose.

The ability of the involved mechanisms to distribute the possible changes to the whole of the flow field in an efficient and smooth way, is surely one of the most important properties of the procedures.

In consideration of the performance and versatility thus demonstrated, together with the new view point presented, the new procedures are expected to be candidates for providing a powerful impulse and a fresh breath into the area of computational fluid dynamics.

Refined for certain physical applications, they may even be considered as an on-board, dynamic means of CFD designs, perhaps paving the way to shape

© Springer International Publishing Switzerland 2016

Z. Mazhar, *Fully Implicit, Coupled Procedures in Computational Fluid Dynamics*,
Fluid Mechanics and Its Applications 115, DOI 10.1007/978-3-319-29895-5_10

changing mechanisms during operation; something which may stress our imagination further.

As it is true for any numerical scheme, it is possible to further enhance the performance of the procedures presented, in various ways. Some suggestions in this regard may be beneficial to enthusiastic developers. Hence before closing, a few comments on further enhancements and recommendations are appropriate.

So far the off-diagonal vectors of the discretized forms of the momentum equations in A_{11} and A_{22} have all been kept as defect vectors in D. It is possible to improve the performance of the procedures by keeping some of these as strength vectors in the L and U matrices. This can be accomplished, for example, by keeping the lower diagonal vectors of A_{11} and A_{22} in L, and the upper diagonal vectors in U. The process is similar to the technique used in FICS-1 and FICS-2 and does not cause any problem. Various tests conducted indicate that some additional ten to fifteen percent savings in computing time can be realized by this improvement. While some more durability can be achieved in this way, the increase in extra storage and formulae may, at times, balance the resulting gains.

A further enhancement can be realized by partially cancelling the effects of the defect vectors by a linear distribution of the effects to the neighboring grid points. If the burden of some more complicated equations is envisaged, another five to ten percent faster convergence rates can be achieved.

Various other ways of replacing the strength and defect vectors may also be considered. For sure, there is no reason why some other ingenious placement may not work better.

Could CFD be any simpler and more enjoyable?

Appendix A
A Critical Survey of Literature—An Adventure into Perfection

The development of computing techniques for the solution of field problems has an interesting history. This development can be truly regarded as an adventure into perfection. The evolution was initiated and supported by valuable contributions of many researchers. The Author of this book feels that the omission of these contributions would render the study somehow incomplete.

The following brief survey is aimed to serve this purpose and perhaps aid the readers seeking more information on various aspects that were not adequately mentioned in the book. The Author tried to do his best in including all of the historical contributions, but if, however any contributions were omitted, it was unintentional and apologizes for such in advance.

The survey is not just a plain, historical list of the contributions. It also includes some useful critical observations and discussions on the strengths and weaknesses regarding these contributions. The Author sincerely believes that the discussion of the weaknesses as well does not hinder the importance of the contributions. On the contrary, a careful assessment of these experiences may assist in the development of more powerful procedures for the benefit of humanity.

The importance of numerical simulation of incompressible fluid flow problems became more understood in parallel with the advent of modern computing machines in the late 60s and early 70s. It was observed that there were two phases that must be dealt with care for such solutions. The first one of these was to develop the most suitable discretization technique to be used in order to convert the set of partial differential equations into a set of simultaneous linear set of equations, which can describe the velocity and pressure fields adequately. When solved with a certain accuracy, these equations would give the required solution to the problem. Depending on the type of discretization, a grid structure was needed to complement the discretization.

The second phase was to develop a suitable equation solver for the resulting set of linear equations. In these early years, finite differences were very well known for a long time since. Later on, the advancement in computing machinery urged researchers to the development of various methods such as Finite Elements, Boundary Elements and Finite Volumes. Despite this, the use of finite difference methodology continued, mainly due to its simplicity and versatility. An important

© Springer International Publishing Switzerland 2016
Z. Mazhar, *Fully Implicit, Coupled Procedures in Computational Fluid Dynamics*,
Fluid Mechanics and Its Applications 115, DOI 10.1007/978-3-319-29895-5

feature of finite differences was that the resulting linear set of equations possesses a well-structured form. This feature enabled the development of more efficient algorithms for the solution of these equations.

The adventure started with the development of upstream differencing methods first proposed by Runchal et al. [40] and studied further by Torrance and Rockett [46] and Raithby and Torrance [37]. Considering the utmost importance of the mass conservation constraint, the staggered grid arrangement was proposed first by Harlow and Welch [19]. In this arrangement, the velocity components were placed exactly on the faces of the control volumes. In this way, the physical meaning of the constraint was reflected more efficiently into the approximations. These methods were analyzed and evaluated thoroughly by Raithby [39] after which they became almost standard in all applications.

The upstream differencing scheme was developed upon realizing that the assumptions made for the variation of dependent functions over a certain region play a vital role in numerical development. This was due to the (ρu) term causing the nonlinearities in the momentum equations in the convection term. It was realized that in these terms, the use of straight averaging of the dependent variables at an interface of a control volume was causing difficulties. This was because the changes of the dependent variables at a grid point were not properly realized at the points downstream of the flow. There appeared some unexpected negative coefficients in places where they should have been physically positive. Using the 'upstream' values instead of straight averages solved the problem.

Mathematically, the problem actually lay in the choice of the linearization of the dependent functions. If it was known that the dependent functions behave in a certain shape, the linearization process must have been chosen in such a way as to reflect this shape. Several schemes were developed for this purpose. The Power-Law Scheme [32], the Hybrid Scheme [34] and the Exponential Scheme (named also as the Upstream Weighted Scheme) [37] were then developed for this purpose.

The Upstream Weighted Scheme, coupled with the staggered grid arrangement proved to be useful remedies for such problems. These have been used successfully until nowadays and they are almost classical.

The solution of the nonlinearity problem was straightforward and could be handled by a suitable iterative approach. However, some fast changes usually appearing in the early stages of flow development was causing divergence in the numerical process. In order to obtain convergence, it was necessary to slow down these changes by the enforcement of some under-relaxation. There was no problem for time dependent flow situations, since the time step used may have been chosen smaller in order to cope with such abrupt changes. To avoid divergence and to provide faster convergence for steady flows, Raithby and Schneider [38] proposed the E-factor formulation. This provided a remedy for the problem.

Following the formulation of the primitive finite difference equations, the next phase of the solution process was the solution of the resulting linear system of equations. Researchers then realized that besides other problems there was an important and crucial problem with the solution of these equations, no matter what discretization technique has been used. This was the velocity-pressure coupling problem. The real difficulties then started to show up.

The use of direct solution techniques such as Gaussian elimination for the solution of the primitive system was possible, but costly due to the sparseness of the matrices involved. The huge amount of storage and computing time required for solution with some available sparse matrix solvers rendered a direct solution completely formidable.

The alternative was to utilize an iterative solver. However, various available iterative solution techniques were not suitable for the purpose. This was due to the violation of certain conditions necessary for the proper functioning and convergence of the methods. The matrices involved were not diagonally dominant, a condition which is strictly necessary for the convergence of an iterative process. In addition, the presence of some zero coefficients in the main diagonals of the matrices involved, rendered an iterative process completely inapplicable. Reordering of the equations in order to bring nonzero coefficients to the main diagonal did not seem favorable either.

In consideration of these difficulties, researchers diverted to a 'divide-and-conquer' approach, named as *segregated solution*. There was no problem in solving the equations for the velocity components separately if the pressure field was given. Some assumed velocity and pressure fields would suffice, which could be corrected afterwards. In addition, an assumed velocity field would cure the nonlinearities as well. This would then be cast into an iterative scheme in which mass and momentum conservation would be satisfied gradually, when and if the process converged.

For this purpose, the velocity and pressure fields need to be corrected. Since the resulting velocity field would not satisfy the mass conservation constraint at an intermediate stage of the process, it needed correction. The corrections started on the intermediate velocity field. It was assumed that the velocity field could be corrected by the assumption that a *pressure correction* exists for this purpose. This involved some assumptions to be made on the velocity field, but nevertheless an 'equation' for a 'pressure correction' field was at hand. The velocity field was then updated using this 'pressure correction' field.

However, this time the momentum equations would not be satisfied since the velocity field satisfying momentum had been changed. If the latest pressure field was 'corrected' to surpass the effect of the assumptions made, it would give a better satisfaction. At first, realizing that the 'pressure correction' field had similar characteristics as that of the pressure field, the 'pressure corrections' might have been assumed as 'pressure differences', hence they were utilized directly to update the pressure field. This resembled, with an old saying, the question 'does the hen come from the egg or does the egg come from the hen?'

Although not worded or pursued, such a correction field was in fact readily available by the 'pressure difference' terms in the momentum equations.

Nevertheless, despite its slow convergence character, the process was successful. An improved satisfaction of the mass and momentum conservation was sought from one overall iteration to another. The iterations thus formed were continued until and if convergence resulted.

The slow convergence of the process led researchers to investigate the reasons. It was decided that the assumptions made in the process were the main culprit. A series of enhancements were then proposed in order to obtain better convergence rates by easing the unrealistic assumptions made. The operations were successful, but of course, with some penalty.

In justifying and appreciating the efforts and assessments of the developments achieved towards the treatment of the problem, a closer examination of the pioneering works of researchers is necessary.

Harlow and Welch [19] proposed one of the earliest methods to account for the velocity-pressure coupling. A Poisson-type equation was derived by combining the mass and momentum equations. Chorin [7, 8] developed two methods, the first being for steady state problems, by trying to adapt the ideas of Harlow and Welch for compressible flows by using an artificial equation of state. While Chorin was trying to apply his artificial compressibility method to transient problems, he came up with a 'velocity correction' idea. Recognizing the fact that the convective and pressure terms in the governing equations can be written as a sum of a quantity with zero divergence and a quantity with zero vorticity; the essential feature of his procedure was to sequentially use the part with zero divergence to calculate an intermediate velocity field which had the correct vorticity but did not satisfy mass conservation, and then to use the part with zero vorticity to correct this intermediate field to ensure mass conservation. Chorin's method comprised the basis of several other procedures, some of them still being used, and therefore is of fundamental importance.

Amsden and Harlow [1], following Chorin's ideas, proposed that a potential function, determined by a Poisson equation, be used for the correction of the velocity divergence without affecting the vorticity of the intermediate velocity field.

Patankar and Spalding [35] developed the well-known SIMPLE (*S*emi-*Im*plicit *P*ressure *L*inked *E*quations) method in which the correct pressure field was written as the sum of an intermediate pressure field and a correction on this pressure. The pressure correction, determined by a Poisson equation, was first used to correct the intermediate velocity field to ensure mass conservation. Then, pressures were updated by simply adding the pressure corrections to the intermediate pressures.

Later, Briley [3] proposed the use of a potential function to update pressure. This potential function would be calculated from a Poisson equation.

In their efforts to examine how quickly convergence is promoted with these methods, Raithby and Schneider [38] came up with new ideas, one of them being the *E*-Factor formulation in which a 'consistent time step' was used for updating pressure. This was actually an under-relaxation process. It helped in suppressing the high velocity and pressure changes that usually appear at the very early stages of flow development. In this way, divergence of the iterative process was avoided. They also proposed the procedure called PULS (*P*ressure *U*pdate by *L*east *S*quares) to determine a pressure distribution which satisfies the momentum equations in a least-squares sense. This method contained Briley's method as a special case.

Patankar [36] derived the SIMPLER method which was a revised form of his original method in which some more realistic assumptions were made in the formulation of the pressure equations. This enhancement was more efficient in updating the pressure field, but required the solution of another Poisson-type equation.

When mathematicians are asked, they immediately recall that the Poisson equation is one of the partial differential equations whose solution is the most difficult. For the pressure field resulting in incompressible flows, the problem becomes even more important. It can be seen clearly from the momentum equations that not the value of pressure at a grid point, but only the amount of 'pressure differences' between the grid points is required to determine the velocity field. This 'relativeness' of the pressure values necessitates that the pressure is let 'free' at the boundaries, that is to say, all boundary conditions be of Neumann type, but perhaps at one 'reference' point at which it is set to some fixed value, say zero.

This physical reality however, reflected itself as a mathematical difficulty in the solution stage. Since all but only one of the linear equations resulting from the discretization of the Poisson equation were weakly diagonally dominant, iterative solution of these equations suffered from very slow convergence. Divergence might even result if strong under-relaxation was not applied. In fact, it was realized that up to 90 % of the computing time required for the solution of the flow problem was spent on the solution of these artificially formed Poisson-like equations.

Realizing this deficiency, researchers started to search for some ways of avoiding the use of a Poisson-type equation. In this regard, Raithby and Schneider [38] proposed the PUMPIN (*P*ressure *U*pdate by *Mu*ltiple *P*ath *In*tegration) method. This method did not require the solution of a Poisson equation. Instead, the pressure field was updated by starting from the reference point where pressure was set to zero and integrate the momentum equations along two paths to each interior pressure point, and then averaging these results to obtain a new pressure field.

Mazhar and Raithby [27] presented NEW-PUMPIN which was a refined version of PUMPIN in which integrations were not only performed along two paths, but along all the possible paths to a pressure point in the solution field. The procedure was based on a binomial weighting technique whose complete pass cost only as much as one pass of the original method. The new method performed much better than the original one.

In an effort to obtain an idea of an ideal one on the emerging procedures, Mazhar [28] examined all of the above described segregated procedures and performed a comprehensive evaluation of their performances in coping with the velocity-pressure coupling problem. He concluded that the NEW-PUMPIN performed best for small-sized problems, but for larger problems, the distance of a single reference point was causing slower convergence rates. This was due to the inability of the mechanism to quickly and efficiently distribute the changes to the whole solution field. The PUMPIN method was also causing problems when obstructions were present in the flow field. In conclusion, the SIMPLER method was found to be more stable and faster. Research was then concentrated on the improvement of various aspects in the segregated solution procedures.

At the time, the iterative procedures like Gauss-Seidel, Successive Over-Relaxation and the Alternating Direction Implicit methods were used to solve the Poisson equations. Realizing the slow convergence of these methods, researchers concentrated on developing better algorithms for solving these equations more efficiently. Based on an original idea of Stone's [44] SIP (*S*trongly *I*mplicit *P*rocedure), Schneider and Zedan [42] came up with a modified version called MSIP whose performance was quite satisfactory. The main idea in these methods was to perform an incomplete decomposition of the form $A = LU + D$ onto the five-diagonal coefficient matrix A resulting from the Poisson equation, followed by successive iterations to overcome the effects of the 'defect matrix' D. In SIP, the five strength vectors in L and U were placed in the same places as that of the vectors of A, resulting in two defect vectors in the matrix D at locations adjacent to the lower and upper triangular diagonals but closer to the main diagonal. Then the effect of these defect vectors was partially cancelled by linearly distributing these effects to the variables at the nearby grid points. In MSIP, two additional strength vectors were employed, one above the lower diagonal of L and one below the upper-diagonal of U. In this way, the places of the defect vectors were shifted one place closer to the main diagonal of D. This was followed by a partial cancellation of the effects of the defect vectors. With these modifications, MSIP proved to be much faster than SIP. Moreover, it was observed that it could withstand the destructive effects of some large aspect ratios of grid sizes in both directions of the grid.

MSIP was a step forward in the improvement of the solution procedures, as expected. In fact, if the idea of employing more strength vectors was continued until the three diagonals close to the main diagonal, this would result in a complete factorization. However, this would be the same as employing a full banded Gaussian elimination solution, requiring full storage of the band and resulting in high cost of computing time. More improvements were also suggested in parallel, by utilizing a nine-point discretization scheme. However, increasing storage requirements was discouraging researchers and efforts in this regard were not continued.

These improvements comprised a leap forward in decreasing the cost in the solution of the linear equation sets, but they had no contribution towards better treatment of the velocity-pressure coupling problem. The reason for slow overall convergence rates and highly confined stability ranges was still not resolved.

Efforts on developing better algorithms for the treatment of the velocity-pressure coupling problem continued in various ways of easing the effects of the assumptions made in the SIMPLE-type methods. In this respect, several enhancements were proposed. Van Doormaal and Raithby [47] suggested SIMPLEC (SIMPLE-*C*onsistent) and Latimer and Pollard [25] presented FIMOSE (*F*ully *I*mplicit *M*ethod for *O*perator-*S*plit *E*quations). Latimer and Pollard presented a comparison of SIMPLER, SIMPLEC and FIMOSE and concluded that the overall performance of the methods with these enhancements are more or less similar, but that SIMPLEC produces relatively faster results.

All of the above enhancements required the solution of two Poisson-type equations, one for pressure and one for pressure correction. Issa [21] suggested PISO (*P*ressure-*I*mplicit by *S*plitting of *O*perators) for correcting pressure once more in each iteration by the use of an additional Poisson equation. This resulted in a predictor-corrector approach in which a corrector for pressure was used two times before proceeding with the next iteration of the solution process. Chow (9) suggested some improvements to PISO, which resulted in some ten percent reduction in computing time as compared with SIMPLER, but admitted that the computation time used by these methods was too long and that further research was needed in improving their speed and economy for practical fluid flow problems.

Although working well, these procedures were poor in dealing with the nonlinearities. In contrast to the reference to *I*mplicitness in their code names, these procedures were of the semi-implicit type. Despite these improvements, segregated solution procedures continued to involve the decoupling of the velocity and pressure variables. Researchers actually were aware that the coupling was the most important property of these variables and should have been preserved for a better and faster interaction between the variables, but the question on how this was to be done, was unanswered. Attention was then focused on some more implicit schemes.

Raithby and Schneider [38] had already pointed out that keeping more of the effects of the coupling between the velocity and pressure fields significantly increases the convergence rates.

The first attempt in this regard was that of Zedan and Schneider [53] in which the velocity components in the mass conservation equation were replaced by their respective values from the momentum equations but keeping pressure terms as unknowns. This resulted in a more implicitly formulated pressure equation which was solved by an SIP-type procedure. Results were promising, but the procedure still was of the segregated type.

Galpin, Van Doormaal and Raithby [15] presented CELS (*C*oupled *E*quation *L*ine *S*olver) in which both of the momentum and mass conservation equations were let coupled on a grid row or column of the flow region. Then the solution was performed based on repeated sweeps in both directions of the grid. This indeed proved to be a more stable algorithm requiring less under-relaxation, as a result that convergence could be obtained for larger values of the *E*-factor. However, since the strong coupling over the whole region was disrupted due to the line-by-line treatment, overall performance of the procedure was almost the same as those of SIMPLER and SIMPLEC. Nevertheless, the findings helped in recognizing that if coupling is preserved more implicitly, more robust algorithms could result.

In a continuing effort for this purpose, Zedan and Schneider [52] provided CSIP (*C*oupled *S*trongly *I*mplicit *P*rocedure) which has been a first attempt in which coupling of the velocity and pressure variables was preserved in the whole domain. The equations for momentum and mass conservation for each point in the solution domain were written together by keeping their primitive form in a 3 × 3 submatrix. That is to say, the variables in the unknown vector was written in the sequence $u_1, v_1, p_1, u_2, v_2, p_2, \ldots..., u_N, v_N, p_N$, and the equations were written sequentially as *u*-eq1, *v*-eq1, mass-eq1, *u*-eq2, *v*-eq2, mass-eq2, …, *u*-eq*N*, *v*-eq*N*, mass-eq*N*. This reordering resulted in a block matrix system whose matrix was five-diagonal, but whose coefficients were those 3×3 submatrices. The solution of this block system was then performed using an MSIP type procedure in which inversions on the 3×3 submatrices were involved. Unfortunately, although the expectations were different, for some reason this procedure performed purely. The overall computational cost was about 4–10 times more, compared to that of SIMPLER.

However, a significant, but not pursued conclusion of this study was that the convergence ranges for the E factor and that of the α parameter used for under-relaxation in the solution of the linear equation sets were further widened than the previous methods.

The above observation was really 'telling something', but unfortunately, in consideration of the poor performances of the methods, research toward fully implicit procedures utilizing the primitive forms of the mass and momentum equations was somehow relaxed and attention was focussed back onto segregated-type procedures.

Meanwhile, the use of the segregated-type procedures continued [49]. Suggestions for improvement of these procedures in some refined aspects have also been made. Chatwani and Turan [5], observing that stronger relaxation in the early stages of flow development may be helpful, proposed a variable under-relaxation factor for use in the solution algorithms. Kim and Ro [23] presented an improvement called MCGS (*M*odified *C*onjugate *G*radient *S*olution) for the solution of the Poisson Equations. A block-correction algorithm was presented by Sathyamurty and Patankar [41]. Kim and Ro [23] proposed their BASIP (*B*lock *A*ided *S*trongly *I*mplicit *P*rocedure) utilizing a combination of SIP and a block solver with some improvement in solution costs.

Studies on improvements in various aspects and comparisons reported by Galpin and Raithby [15], were continued by Theodossiou and Soussa [45], Connell and Stow [9], Henau et al. [20], Deng et al. [13], Wen and Ingham [50] and Choi and Yo [6].

It was not until 1995 that attention was focused back onto the fully coupled solution of the momentum and mass conservation equations in their primitive forms.

In an attempt to uncover the mystery in why the approach presented by Zedan and Schneider [52] performed so purely, Güngörmüş [17], using similar mechanisms as those used in CSIP, reported similar observations.

Hanby et al. [18] tried the use of the QMR (*Q*uasi *M*inimal *R*esidual) algorithm of Freund and Nachtigal [13] for the solution of the equations in their primitive forms. In parallel to the results obtained by Zedan and Schneider [52] and Güngörmüş [17], the fully coupled approach proved to be more effective in dealing with the nonlinearities, but poor in the overall convergence characteristics.

In the meantime, an attempt was made by Lage [24] in an effort towards improvement of Poisson solvers by performing an adaptive optimization of the MSIP procedure with some success. Sheen [43] applied the PCGS algorithm, reporting some good convergence and stability but lower asymptotic convergence rates.

It was obvious by the time, but unfortunately not openly worded, that the failure of the fully coupled approaches with an effort to solve the primitive equations efficiently, was due to the solution procedures themselves. The question was: why are such procedures so slow and costly, despite the improvements in stability?

The culprit to this dilemma was discovered in an attempt by Mazhar [29] in which the primitive equations were solved by an incomplete decomposition technique applied to the block primitive matrix, similar to that used in the SIP and MSIP. The main problem was the following:

In the previous studies, the block matrix was formulated by rewriting the equations in a pointwise way. This was quite all right, but what happened later on in the solution process was of utmost importance. With this type of ordering, the main diagonal coefficients of the momentum equations were becoming less effective than those at the off diagonals. The effects of the diagonal coefficients of the neighboring points were scattered and distributed to the off diagonals of the block matrix, thus worsening the diagonal dominancy of the matrix. The result, inevitably, was slower convergence, which had to be partially treated with strong relaxation.

Mazhar's formulation of the block system was quite different. The two momentum conservation equations were written as two consecutive blocks, followed by the equations of mass conservation for all points in the solution field. That is to say, the equations were retained in their original, primitive form, as commanded by the governing partial differential equations. Then it was discovered that an incomplete decomposition works for this block matrix, despite the fact that one of the main diagonal block sub-matrices referring to pressure was completely null. With the special incomplete decomposition technique used, nonzero coefficients were able to be created in all of the main diagonals of the strength matrices L and U, even at the places corresponding to the null matrix for pressure. More importantly, with the mechanism involved in this decomposition, the effects of the dominant diagonal coefficients of the momentum equations, together with the effects of the coefficients of pressure terms and those of mass conservation were all successfully transported into the strength vectors of L and U. Moreover, the effect of the defect vectors was much less than those of the strength vectors.

The results were astonishing. Convergence rates of this fully implicit coupled procedure were much more than anticipated. On the other hand, the number of relaxation parameters was reduced to two and the intervals of convergent relaxation parameters were broadly widened. To the benefit, the procedure did not need the solution of a Poisson-type equation and the formulation was extremely easy.

The procedure was adapted on the existing codes of some other independent reasearchers who at first found the claims to be 'too good to be true'. The procedure was tested on various problems for about a year with the conclusion that it was in fact at least twice as fast and much more stable compared to SIMPLER. This was the procedure named BIP in this book.

With the encouragement of the performance of the procedure, research continued toward improvement of the mechanism involved. An enhancement to this procedure called BIPEN was formulated by Mazhar [30] which proved to be about 40 % faster and more durable than the BIP.

These observations paved the way for further enhancements. It was noticed that various other ways were possible in the selection of places of the strength vectors in the L and U matrices. FICS-1 and FICS-2 [31] use similar mechanisms but include even broader enhancements.

The procedure FICS-1 converged much faster than the former ones and the range of convergent relaxation parameters was much widened. The climax of these was FICS-2, which proved to be about twenty times faster than any of the segregated-type ones. In fact, solutions could be obtained with about 5 % of the cost of the SIMPLER-type procedures. The method proved to be convergent even without any relaxation. The elimination of the need for Poisson-type solutions and the simplicity in the formulation and solution stages were the extra benefits of these procedures. The broadly expanded ranges of optimal relaxation parameters were a reference to the high stability of the procedures.

A further enhancement, which was surprisingly the simplest of the family of procedures, named as SICS, was developed by Mazhar. This procedure was not as strongly robust as FICS-2, but it had several, important advantages over FICS-2. Pointwise applicability, simplicity of formulations and elimination of the need for extra storage were some of the extra benefits.

Despite these developments, the SIMPLER-type procedures are still widely used in the simulation of fluid flow problems. Efforts even continue on enhancing the solution of the Poisson-type equations involved in the segregated-type procedures. The latest of these efforts was reported by Chalhub et al. [4] in which a semi-analytical formulation was proposed.

Promisingly, on the other hand, coupled algorithms have recently regained some more attention. Darwish et al. [11] derived a 'pressure equation' similar to a segregated SIMPLE algorithm by considering the primitive form of the equations. Mangani et al. [26], using a SIMPLE type algorithm in a finite volume approach, reported that implicit block coupling of pressure and velocity variables leads to faster convergence compared to classical fully uncoupled ones. Ashrafizadeh et al. [2] reported similar conclusions as well. The work of Darwish et al. [12] involves extracting an implicit equation for pressure following a segregated mass conservation based algorithm in a finite volume scheme with which 1.3 to 4.6 times faster convergences are reported.

Although substantial improvement in robustness has been achieved with these procedures, the overall performances of the proposals are overshadowed by complicated algorithms. The expected high performances were not realized, perhaps basically because the naïve full implicitness of the coupled primitive equations was not utilized favorably. The proposed procedures were either carrying some segregation or, even if the coupled forms were used for some implicitness, they were still pressure-based.

It has been obvious by now that the common approach of trying to obtain an explicit equation for pressure somehow dissolves the strong coupling inherently. Hence, it has become clearly understood that the strong coupling in the primitive equations is not a disadvantage, but rather it is a property that must be carefully preserved and advantageously utilized.

The Author of this book feels that the development of the segregated-type or pressure-based procedures climbed to a saturation, with no further significant advance. Therefore, much attention must be focused on the latest coupled fully implicit treatment of the velocity-pressure coupling problem.

Appendix B
Segregated Solution Procedures: SIMPLE and SIMPLER

The segregated type procedures have been widely used for over four decades, despite their slow convergence rates and stability problems. Due to their historical importance, two of these procedures are presented here, for the sake of completeness and for the interest of the reader.

The discretization of the governing partial differential equations as described in Chap. 4 resulted in three linear(ized) equations for each grid point interior to the solution domain under consideration. For each nodal point P, equations of the following form were obtained:

$$A_S^u u_S + A_W^u u_W + A_P^u u_P + A_E^u u_E + A_N^u u_N + A_P^{pu} p_P + A_E^{pu} p_E = b_P^u \tag{B.1}$$

$$A_S^v v_S + A_W^v v_W + A_P^v v_P + A_E^v v_E + A_N^v v_N + A_P^{pv} p_P + A_N^{pv} p_N = b_P^v \tag{B.2}$$

$$A_W^{Cu} u_W + A_P^{Cu} u_P + A_S^{Cv} v_S + A_P^{Cv} v_P = 0 \tag{B.3}$$

We note that Eqs. B.1 and B.2 are the same as Eqs. 4.32 and 4.39, respectively. However, Eq. B.3 differs from Eq. 4.44 in that the right hand side is set to zero. This is because the boundary conditions have not been applied onto the equation.

The nonlinearity of the differential equations is reflected in the difference equations by the coefficients which themselves depend on the unknowns.

The earliest procedures for solving these equations which were economically viable for most applications were proposed by Patankar and Spalding [35] and Briley [3]. Many variations and enhancements to these ideas have been proposed in due course as described in Appendix A.

All of these procedures work in the following fashion:

Step 1. Guess a preliminary pressure field p^*, or use pressure values available from a previous solution step.

Step 2. Solve Eqs. (B.1–B.2) using this p^* field to obtain a new velocity field. Denote these velocities by u^* and v^*.

Step 3. Correct u^* and v^* velocities to u^c and v^c such that u^c and v^c satisfy Eq. B.3.

Step 4. Update coefficients.

© Springer International Publishing Switzerland 2016

Z. Mazhar, *Fully Implicit, Coupled Procedures in Computational Fluid Dynamics*, Fluid Mechanics and Its Applications 115, DOI 10.1007/978-3-319-29895-5

Step 5. Update pressure field.
Step 6. Repeat Steps 2–5 until convergence.

A complete cycle consisting of Steps 1–6 is called an *overall iteration*.

Details of Steps 1 and 2: Solution of the momentum equations

To begin with, it is assumed that some knowledge about the velocity and pressure fields is at hand. If not, some appropriate initialization of the velocity and pressure fields is necessary. Although not necessary, for most problems, initialization to zero may be appropriate.

Step 1 of the segregated process is to *guess* a pressure field. For the first overall iteration, the pressures may be assumed to be zero. In the later stages, the most recent pressure values already available from a previous outer iteration are used.

In Step 2, these pressures are used in the two momentum equations so that the following equations result:

$$A_S^u u_S^* + A_W^u u_W^* + A_P^u u_P^* + A_E^u u_E^* + A_N^u u_N^* + A_P^{pu} p_P^* + A_E^{pu} p_E^* = b_P^u \tag{B.4}$$

$$A_S^v v_S^* + A_W^v v_W^* + A_P^v v_P^* + A_E^v v_E^* + A_N^v v_N^* + A_P^{pv} p_P^* + A_N^{pv} p_N^* = b_P^v \tag{B.5}$$

Assembled in a matrix form, these equations become

$$[A^u]\{u^*\} = \left\{b^{u^*}\right\} \tag{B.6}$$

and

$$[A^v]\{v^*\} = \left\{b^{v^*}\right\} \tag{B.7}$$

These matrix systems are shown explicitly in Figs. B.1 and B.2.

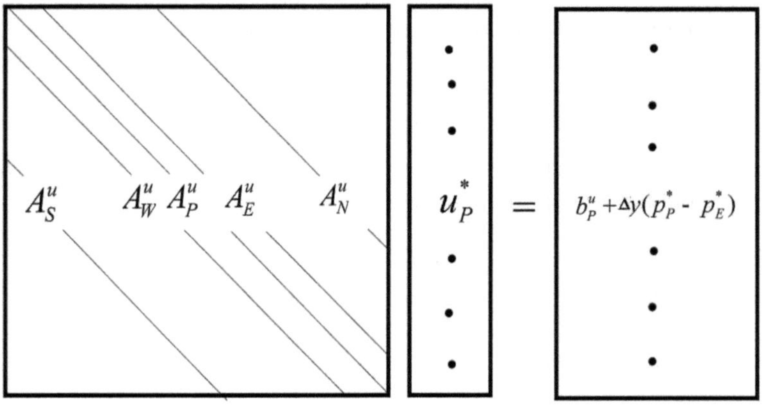

Fig. B.1 Matrix system for *u* velocity in segregated solution

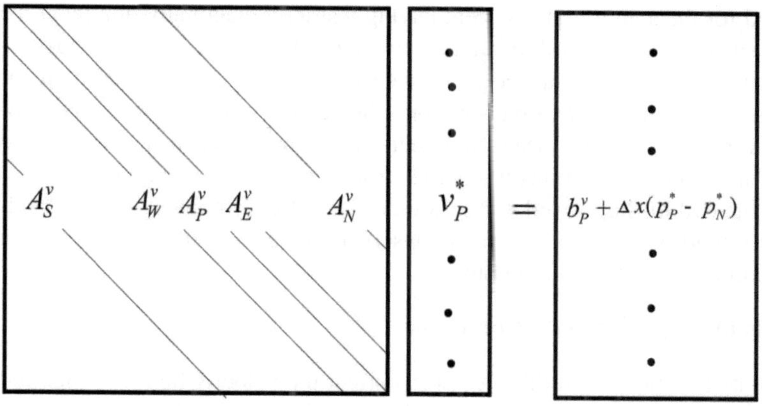

Fig. B.2 Matrix system for v velocity in segregated solution

The two coefficient matrices are *five-diagonal*, that is to say, all the coefficients, except those on the indicated five diagonals are zeros.

Besides these matrices being highly sparse, it is more important to note the following observation: From the formulation of the coefficients it follows that

$$|A_P^u| = \frac{1+E}{E}\sum |A_{nb}^u| \tag{B.8}$$

and

$$|A_P^v| = \frac{1+E}{E}\sum |A_{nb}^v| \tag{B.9}$$

in which A_{nb} denotes the coefficients of the four neighboring points of P. Since $(1+E)/E > 1$ for all $E > 0$ and $\lim_{E\to\infty}\frac{1+E}{E} = 1$, it follows that

$$|A_P^u| \geq \sum |A_{nb}^u| \tag{B.10}$$

and

$$|A_P^v| \geq \sum |A_{nb}^v| \tag{B.11}$$

With a finite value of E, the matrices are strictly[1] diagonally dominant, while as $E \to \infty$ they become loosely diagonally dominant. Convergence of an iterative

[1]Matrices whose coefficients satisfy the property in Eqs. (B.10, B.11) are called *diagonally dominant* in the literature. The case in which only strict inequality is valid is defined by Varga (48) as *strictly diagonally dominant*. On the other hand, Jennings (24) uses the term *loosely diagonally dominant* for the case in which only equality holds. For a loosely dominant matrix, the inequality must hold for at least one row, since otherwise a unique solution to the system with such a matrix does not exist.

method for the solution of a system having a strictly diagonally dominant matrix is faster since the diagonal dominance is stronger. However, if large values of E are used, the diagonal dominance is weakened, with the consequence that the convergence of a suitable iterative solution scheme slows down considerably. We note that a very large E value means no under-relaxation, but in such a case, the solution of the momentum equations will suffer from very slow convergence.

At the end of Step 2, unless the solution has already converged, the resulting *-velocity field will not satisfy mass conservation, because the p^* field was only an approximation to the *true* pressure.

Details of Step 3:Velocity Correction

The primary objective in this step is to correct the velocity field obtained in Step 2 in such a way that the corrected velocity field satisfies the mass conservation constraint. All the available proposals perform this task by postulating the existence of a function $p^{'}$ which satisfies relations of the form

$$u_P^c = u_P^* + f_P^u(p_P^{'} - p_E^{'}) \tag{B.12}$$

and

$$v_P^c = v_P^* + f_P^v(p_P^{'} - p_N^{'}) \tag{B.13}$$

where various f_P's lead to various methods. Here the u and v values superscripted by c denote the corrected u and v values. Substituting Eqs. (B.12–B.13) into the mass conservation equation (B.3) results in a finite difference analog of a Poisson-type equation for $p^{'}$ of the form

$$\left[A^{p'}\right]\{p'\} = \left\{b^{p'}\right\} \tag{B.14}$$

Since $p^{'}$ is assumed to be of the same character as p, boundary conditions for p can be applied to $p^{'}$ as well.

The SIMPLE procedure

Patankar [33] suggested the following scheme for the velocity correction: Consider Eqs. (B.1 and B.2) in which u's and v's are replaced by u^c's and v^c's to write

$$A_S^u u_S^c + A_W^u u_W^c + A_P^u u_P^c + A_E^u u_E^c + A_N^u u_N^c + A_P^{pu} p_P + A_E^{pu} p_E = b_P^u \tag{B.15}$$

$$A_S^v v_S^c + A_W^v v_W^c + A_P^v v_P^c + A_E^v v_E^c + A_N^v v_N^c + A_P^{pv} p_P + A_N^{pv} p_N = b_P^v \tag{B.16}$$

Subtracting Eqs. (B.4 and B.5) from Eqs. (B.15 and B.16) respectively, and rearranging, gives

$$A_P^u(u_P^c - u_P^*) = -A_S^u(u_S^c - u_S^*) - A_W^u(u_W^c - u_W^*) - A_E^u(u_E^c - u_E^*) - A_N^u(u_N^c - u_N^*) \\ - A_P^{pu}(p_P - p_P^*) - A_E^{pu}(p_E - p_E^*) \tag{B.17}$$

and

$$A_P^v(v_P^c - v_P^*) = -A_S^v(v_S^c - v_S^*) - A_W^v(v_W^c - v_W^*) - A_E^v(v_E^c - v_E^*) - A_N^v(v_N^c - v_N^*) \\ - A_P^{pv}(p_P - p_P^*) - A_N^{pv}(p_N - p_N^*) \tag{B.18}$$

Patankar [33] then 'drops' the terms containing velocity differences in the right hand side of Eqs. (B.17 and B.18). Upon defining a p' by

$$p = p^* + p' \tag{B.19}$$

Eqs. (B.12 and B.13) become, after using Eqs. (4.33e–f) and (4.40e–f) and rearranging,

$$u_P^c = u_P^* + (p_E' - p_P')\Delta y / A_P^u \tag{B.20}$$

and

$$v_P^c = v_P^* + (p_N' - p_P')\Delta x / A_P^v \tag{B.21}$$

For a mass control volume around the point P, similar equations can now be written for u_e^c, u_w^c, u_n^c and u_s^c as follows:

$$u_e^c = u_e^* + (p_P' - p_E')\Delta y / A_e^u \tag{B.22}$$

$$u_w^c = u_w^* + (p_W' - p_P')\Delta y / A_w^u \tag{B.23}$$

$$v_n^c = v_n^* + (p_P' - p_N')\Delta x / A_n^v \tag{B.24}$$

and

$$v_s^c = v_s^* + (p_S' - p_P')\Delta x / A_s^v \tag{B.25}$$

Substitution of these values into Eq. (B.3), using $A_W^{Cu} = -\Delta y$, $A_W^{Cu} = \Delta y$, $A_S^{Cv} = -\Delta x$, $A_W^{Cv} = \Delta x$ and rearranging yields

$$\begin{aligned}&\left[\Delta y^2/A_e^u+\Delta y^2/A_w^u+\Delta x^2/A_n^v+\Delta x^2/A_s^v\right]p_P'\\&=(\Delta y^2/A_e^u)p_E'+(\Delta y^2/A_w^u)p_W'+(\Delta x^2/A_n^v)p_N'+(\Delta x^2/A_s^v)p_S'\\&-\left[(u_e^*-u_w^*)\Delta y+(v_n^*-v_s^*)\Delta x\right]\end{aligned}\tag{B.26}$$

Similar equations are then written for all other mass control volumes. Once these equations are assembled into a matrix form as

$$\left[A^{p'}\right]\{p'\}=\left\{b^{p'}\right\}\tag{B.27}$$

they can be solved to obtain the p' field. These are then used in Eqs. (B.20 and B.21) to correct the velocity field.

In the SIMPLE method, Steps 4 and 5 are combined by using Eq. (B.19) to update pressure.

Patankar discussed the implications of the 'approximations' that have been made by dropping the terms involving velocity differences and noted that it has no effect on the converged solution. If and when 'convergence' is reached, u^c's will approach u^*'s, and the mass source of the u^*, v^* field will be zero, therefore giving a pressure correction field of practically zero. In this case it would not make any difference if any approximations are made or not. If the process converges, it converges to the correct velocity and pressure fields.

This was quite meaningful. However, Patankar [36], in his own words, admits that 'the approximations made lead to rather exaggerated pressure corrections, and hence under-relaxation becomes essential. Since the influence of the neighbor-point velocity corrections is removed from the velocity-correction formula, the pressure correction has the entire burden of correcting the velocities, and this results in a rather severe pressure-correction field. In most cases, it is reasonable to suppose that the pressure-correction equation does a fairly good job of correcting the velocities, but a rather poor job of correcting pressure'.

To improve the slow convergence rate of the algorithm he then proposed a revised version of the algorithm, named SIMPLER, as follows:

The SIMPLER Procedure

Once the velocity field has been corrected in Step 3 and new coefficients have been calculated in Step 4, these values are resubstituted into the two momentum equations to obtain

$$u_P^c=u_P^{c*}+(p_P-p_E)\Delta y/A_P^u\tag{B.28}$$

and

$$v_P^c=v_P^{c*}+(p_P-p_N)\Delta x/A_P^v\tag{B.29}$$

in which

$$u_P^{c^*} = -\left[A_e^u u_E^c + A_w^u u_W^c + A_n^u u_N^c + A_s^u u_S^c\right]/A_P^u \tag{B.30}$$

and

$$v_P^{c^*} = -\left[A_e^v v_E^c + A_w^v v_W^c + A_n^v v_N^c - A_s^v v_S^c\right]/A_P^v \tag{B.31}$$

Equations of the form (B.30 and B.31) can be written for the u and v velocities on the face of a mass control volume around a point P as follows:

$$u_e^c = u_e^{c*} + (p_P - p_E)\Delta y/A_e^u \tag{B.32}$$

$$u_w^c = u_w^{c*} + (p_W - p_P)\Delta y/A_w^u \tag{B.33}$$

$$v_n^c = v_n^{c*} + (p_P - p_N)\Delta x/A_n^v \tag{B.34}$$

$$v_s^c = v_s^{c*} + (p_S - p_P)\Delta x/A_s^v \tag{B.35}$$

Substituting Eqs. (B.32–B.35) into the mass conservation equation (B.3), using $A_W^{Cu} = -\Delta y$, $A_W^{Cu} = \Delta y$, $A_S^{Cv} = -\Delta x$, $A_W^{Cv} = \Delta x$ and rearranging gives, for a pressure point P:

$$\begin{aligned}
&\left[\Delta y^2/A_e^u + \Delta y^2/A_w^u + \Delta x^2/A_n^v + \Delta x^2/A_s^v\right]p_P \\
&= (\Delta y^2/A_e^u)p_E + (\Delta y^2/A_w^u)p_W - (\Delta x^2/A_n^v)p_N + (\Delta x^2/A_s^v)p_S \\
&\quad - \left[(u_e^{c*} - u_w^{c*})\Delta y + (v_n^{c*} - v_s^{c*})\Delta x\right]
\end{aligned} \tag{B.36}$$

Similar equations for each pressure point P is written and then assembled into a matrix form as

$$[A^p]\{p\} = \{b^p\} \tag{B.37}$$

Eq. (B.37), which is of Poisson type, can now be solved to obtain the 'new' pressure field.

The matrices for pressure and pressure correction have similar properties, although their right hand sides are different. The boundary conditions of pressure and pressure correction are of zero Neumann type and must be applied to all equations for the grid points neighboring a boundary.

Furthermore, it can be seen from Eqs. (B.27) and (B.37), that the involved matrices satisfy the relation

$$|A_P| = \sum_{nb} |A_{nb}| \tag{B.38}$$

This concludes that the matrices are loosely diagonally dominant, with only one exception: At only one reference point, pressure or pressure correction may be set to a fixed value. This property poses a real challenge to any possible linear equation solver to be used for the solution of the pressure or pressure correction field.

In the early stages of development, the resulting matrices were solved by using either the ADI (*A*lternating *D*irection *I*mplicit), Gauss-Seidel or the SOR (*S*uccessive *Over-relaxation*) methods. These methods proved to be very slow, especially for the pressure and pressure correction. Some gain in speed could be achieved by applying some considerable over-relaxation, but still the speed was not satisfactory.

Researchers, noting the special form of the matrices involved in these equations, tried some new ideas on developing faster methods for these systems. For an appreciation of the fruitful efforts of these researchers and as an extra aid to the reader, we present these methods below.

Method SIP (*S*trongly*I*mplicit*P*rocedure)

It was noted above that the resulting system of equations in the procedures described involves a five-diagonal matrix. Let such a system be denoted by

$$AX = r \tag{B.39}$$

whose graphical form is shown in (Fig. B.3):

The idea of strongly implicitness was first proposed by Stone [44]. Later on, Schneider and Zedan [42] proposed a modified version of Stone's method. For the sake of completeness and a better understanding of the basic ideas, the SIP of Stone will be presented first, followed by Schnedier and Zedan's MSIP.

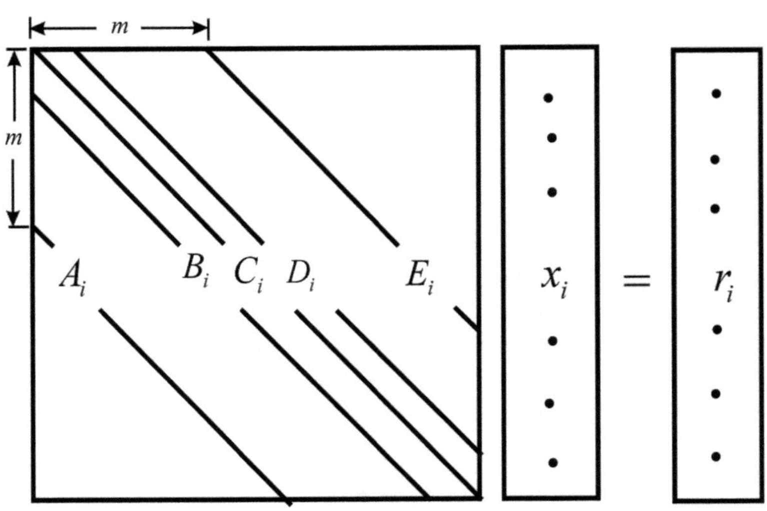

Fig. B.3 Five-diagonal matrix system for SIP

The Author of this book feels obliged to these pioneers for their fruitful ideas in respect that some inspiration, perhaps invoked the development of the procedures presented in the book.

The basic idea in these methods is to modify the given matrix A to a matrix $A + D$, such that the modified matrix can be factored into a lower triangular matrix L and an upper triangular matrix U in less computational effort and lower computer storage than required to factor the original matrix.

For this purpose, Stone noticed that when the two matrices L and U of the forms

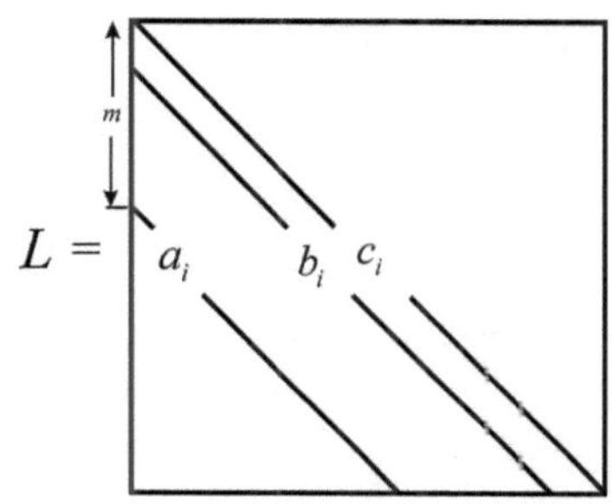

and

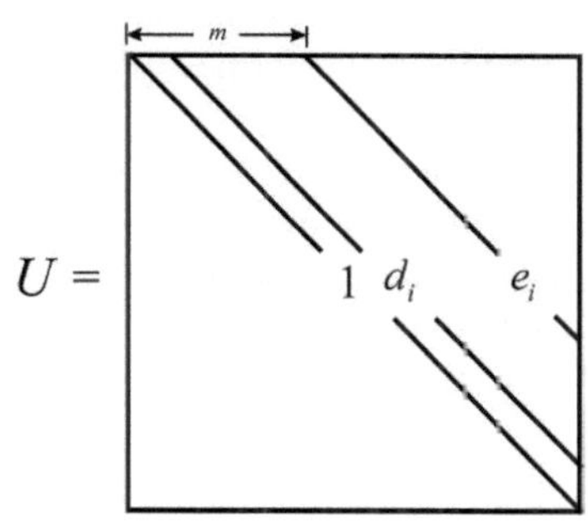

are multiplied, a matrix $A + D$ results which is of the form

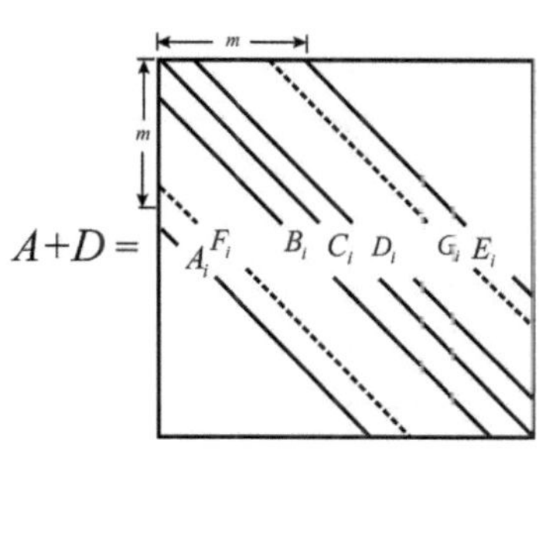

Multiplying L and U and equating the corresponding coefficients result in

$$a_i = A_i \tag{B.40a}$$

$$a_i d_{i-m} = F_i \tag{B.40b}$$

$$b_i = B_i \tag{B.40c}$$

$$a_i e_{i-m} + b_i d_{i-1} + c_i = C_i \tag{B.40d}$$

$$c_i d_i = D_i \tag{B.40e}$$

$$b_i e_{i-1} = G_i \tag{B.40f}$$

$$b_i e_i = E_i \tag{B.40g}$$

As Stone points out, however, it is not possible to 'choose' the L and U matrices such that the modified matrix $A + D$ equals A. Assuming that A_i, B_i, C_i, D_i and E_i are equal to the corresponding terms of the original matrix A, the coefficients of L and U can be calculated by using Eq. (B.40). The additional terms F_i and G_i are then accepted as their nonzero values.

As though not so done by Stone, the vectors in the matrix L and U could be named as strength vectors, and those of D the defect vectors.

It would then be sufficient if a partial cancellation of the adverse effect of the coefficients of the defect vectors is performed. For this purpose, the numerical molecule in Fig. B.4 is considered:

The modified matrix $A + B$ has nonzeros not only at the point i and its four neighboring points, but also at the points $i - m + 1$ and $i + m - 1$. To partially cancel the effect of the variable x at these points, approximately equal terms can be subtracted. Stone writes Taylor Series approximations for x_{i-m+1} and x_{i+m-1} as follows:

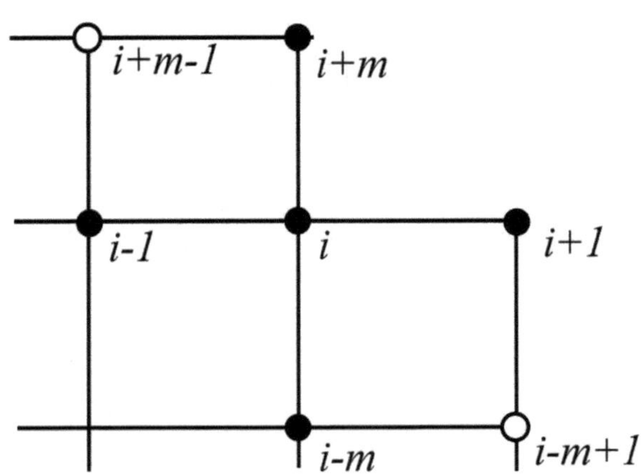

Fig. B.4 Numerical molecule for the Strongly Implicit Procedure

$$x_{i-m+1} = -x_i + x_{i+1} + x_{i-m} \tag{B.41}$$

$$x_{i+m-1} = -x_i + x_{i-1} + x_{i-m} \tag{B.42}$$

He further notes that it would be beneficial to multiply the right hand sides of these equations by an α, where $0<\alpha<1$, as

$$x_{i-m+1} = \alpha[-x_i + x_{i+1} + x_{i-m}] \tag{B.41}$$

$$x_{i+m-1} = \alpha[-x_i + x_{i-1} + x_{i+m}] \tag{B.42}$$

Now, multiplying the modified matrix $A + B$ by the vector X and equating to the right hand side vector r, gives:

$$\begin{aligned} A_i x_{i-m} + B_i x_{i-1} + C_i x_i + D_i x_{i+1} + E_i x_{i+1} \\ + \mathrm{F}_i x_{i-m+1} \\ + \mathrm{G}_i x_{i+m-1} = r_i \end{aligned} \tag{B.43}$$

Now, partial cancellation can be introduced if equivalent values of x_{i-m+1} and x_{i+m-1} are subtracted from Eq. B.43 using Eqs. B.41 and B.42:

$$\begin{aligned} A_i x_{i-m} + B_i x_{i-1} + C_i x_i + D_i x_{i+1} + E_i x_{i+1} \\ + \mathrm{F}_i[x_{i-m+1} - \alpha(-x_i + x_{i+1} + x_{i-m})] \\ + \mathrm{G}_i[x_{i+m-1} - \alpha(-x_i + x_{i-1} + x_{i+m})] = r_i \end{aligned} \tag{B.44}$$

Collecting like terms gives

$$\begin{aligned} (A_i - \alpha \mathrm{F}_i)x_{i-m} + (B_i - \alpha G_i)x_{i-1} + (C_i - \alpha F_i + \alpha G_i)x_i \\ + (D_i - \alpha F_i)x_{i+1} + (E_i - \alpha G_i)x_{i+1} \\ + \mathrm{F}_i x_{i-m+1} \\ + \mathrm{G}_i x_{i+m-1} = r_i \end{aligned} \tag{B.45}$$

Hence Eq. (B.40) should be modified to obtain

$$a_i = A_i - \alpha F_i \tag{B.44a}$$

$$a_i d_{i-m} = F_i \tag{B.44b}$$

$$b_i = B_i - \alpha G_i \tag{B.44c}$$

$$a_i e_{i-m} + b_i d_{i-1} + c_i = C_i + \alpha F_i + \alpha G_i \tag{B.44d}$$

$$c_i d_i = D_i - \alpha F_i \tag{B.44e}$$

$$b_i e_{i-1} = G_i \tag{B.44f}$$

$$b_i e_i = E_i - \alpha G_i \tag{B.44g}$$

Since F_i and G_i are given explicitly from Eqs. (B.44b) and (B.44f) respectively, the remaining equations can be updated to give

$$a_i = A_i - \alpha a_i d_{i-m} \tag{B.45a}$$

$$b_i = B_i - \alpha b_i e_{i-1} \tag{B.45c}$$

$$a_i e_{i-m} + b_i d_{i-1} + c_i = C_i + \alpha a_i d_{i-m} + \alpha b_i e_{i-1} \tag{B45d}$$

$$c_i d_i = D_i - \alpha a_i d_{i-m} \tag{B.45e}$$

$$b_i e_i = E_i - \alpha b_i e_{i-1} \tag{B.45g}$$

Further arrangement of these equations gives

$$a_i = A_i/(1 + \alpha d_{i-m}) \tag{B.46a}$$

$$b_i = B_i/(1 + \alpha e_{i-1}) \tag{B.46c}$$

$$c_i = C_i + a_i(\alpha d_{i-m} - e_{i-m}) + b_i(\alpha e_{i-1} - d_{i-1}) \tag{B.46d}$$

$$d_i = (D_i - \alpha a_i d_{i-m})/c_i \tag{B.46e}$$

$$e_i = (E_i - \alpha b_i e_{i-1})/b_i \tag{B.46g}$$

The above equations can be used to calculate the coefficients of the strength vectors in L and U matrices, which will be used in the forward and backward substitutions in the following iteration sequence.

THE ITERATIVE PROCEDURE

Adding DX to both sides of Eq. (B.39), one may obtain

$$(A + D)X = r + DX \tag{B.47}$$

Adding and subtracting AX to the right hand side of Eq. (B.47) gives

$$(A + D)X = (A + D)X - (AX - r) \tag{B.48}$$

Since the left hand side is now factored into L and U, Eq. (B.48) can be efficiently solved if the right hand side is known. For this, the following iteration scheme can be used:

$$(A+D)X^{(k+1)} = (A+D)X^{(k)} - (AX^{(k)} - r), \tag{B.49}$$

where k denotes iteration level. Defining a difference vector and a residual vector by

$$\delta^{(k+1)} = X^{(k+1)} - X^{(k)} \tag{B.50}$$

and

$$R^{(k)} = r - AX^{(k)}, \tag{B.51}$$

Eq. (B.49) becomes

$$(A+D)\delta^{(k+1)} = R^{(k)} \tag{B.52}$$

Replacing $A + D$ by LU in Eq. (B.52) leads to

$$LU\delta^{(k+1)} = R^{(k)} \tag{B.53}$$

which can be solved in two steps by

Step 1. Solve $L\gamma^{(k)} = R^{(k)}$ (lower solve or forward substitution)
Step 2. Solve $U\delta^{(k+1)} = \gamma^{(k)}$ (upper solve or backward splution)

The last step in one iteration then is to update X by using Eq. (B.50) as

$$X^{(k+1)} = X^{(k)} + \delta^{(k+1)} \tag{B.54}$$

It should be reminded here that the incomplete factorization is performed once and for all and it is followed by a series of lower and upper triangular matrix solutions, until convergence.

Method*M*SIP (*M*odified*S*trongly*I*mplicit Procedure)

The SIP method proved to be a very efficient solution strategy for the type of matrix systems of interest, but it was discovered later on that its performance decreased when the grid structure beared widely varying grid aspect rations.

To combat this deficiency, Schneider and Zedan [42], noting that the strong asymmetric influence of the additional terms would require the renumbering of the grid system after every iteration, considered the numerical molecule shown in Fig. B.5.

The numerical molecule was for a 9-point formulation. Schneider and Zedan, while working on a 9-point finite difference formulation for a better approximation, discovered that the resulting formulations can also be used for a 5-point formulation as a special case. We will shortly see how this is done.

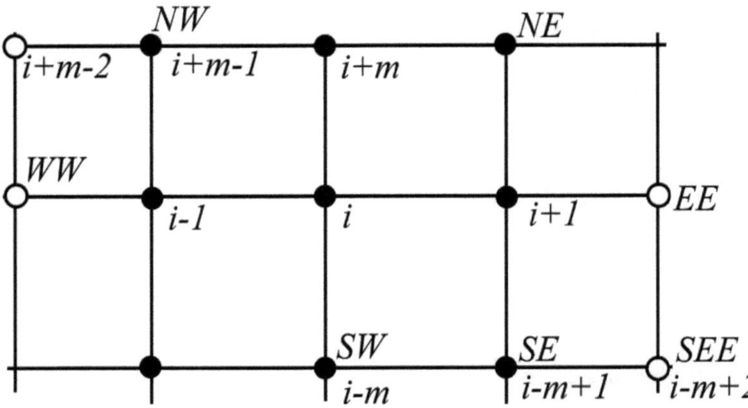

Fig. B.5 Numerical molecule for the 9-point Modified Strongly Implicit Procedure

The corresponding L and U matrices are now given by

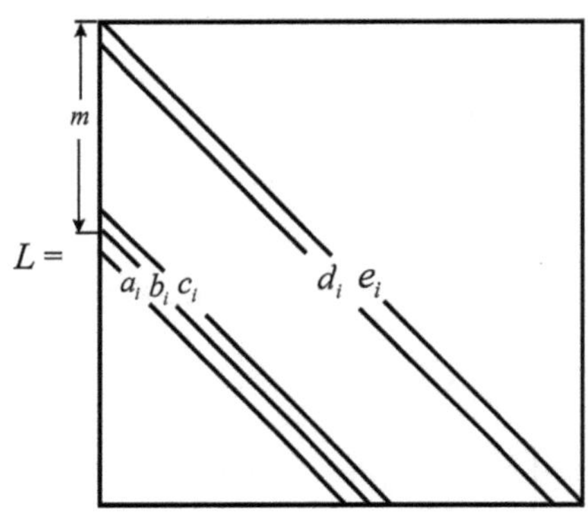

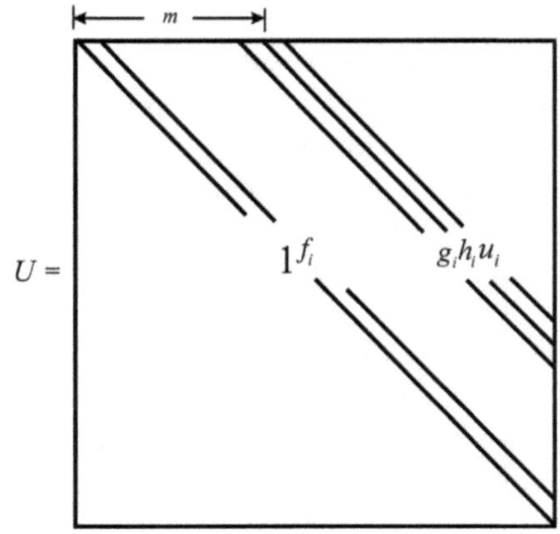

whereas the modified matrix $(A + D)$ is of the form

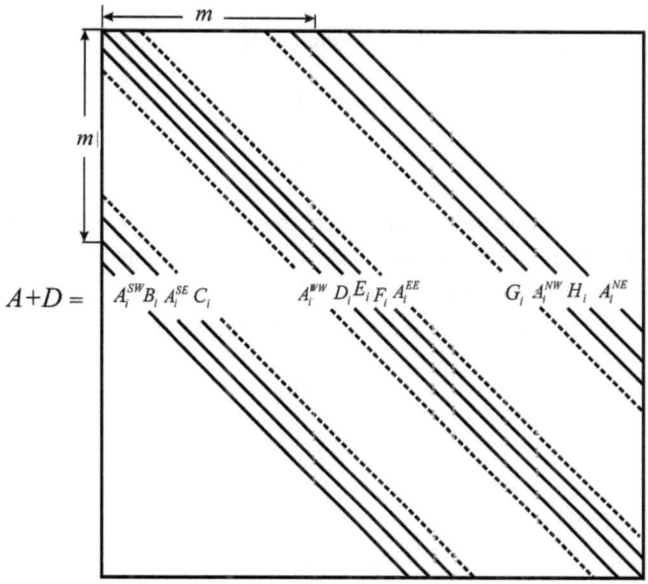

The diagonal vectors shown as solid lines in the matrix $(A + D)$ correspond to the nine coefficients of the original matrix A. The dotted ones are the defect vectors regarding the four corner nodes of the numerical molecule.

Multiplying L and U and equating the corresponding coefficients with those in $(A + D)$ results in

$$a_i = A_i^{SW} \tag{B.55a}$$

$$a_i f_{i-m-1} + b_i = B_i \tag{B.55b}$$

$$b_i f_{i-m} + c_i = A_i^{SE} \tag{B.55c}$$

$$c_i f_{i-m+1\,i} = C_i \tag{B.55d}$$

$$a_i g_{i-m-1} = A_i^{WW} \tag{B.55e}$$

$$a_i h_{i-m-1} + b_i g_{i-m} + d_i = D_i \tag{B.55f}$$

$$a_i u_{i-m-2} + b_i h_{i-m} + c_i g_{i-m+1} + d_i f_{i-1} + e_i = E_i \tag{B.55g}$$

$$b_i u_{i-m} + c_i h_{i-m+1} + e_i f_i = F_i \tag{B.55h}$$

$$e_i u_{i-m+1} = A_i^{EE} \tag{B.55i}$$

$$d_i g_{i-1} = G_i \tag{B.55j}$$

$$d_i h_{i-1} + e_i g_i = A_i^{NW} \tag{B.55k}$$

$$d_i u_{i-1} + e_i h_i = H_i \tag{B.55l}$$

$$e_i u_i = A_i^{NE} \tag{B.55m}$$

For a 5-point formulation, Schneider and Zedan then adopt the numerical molecule shown in Fig. B.6, in which case A_i^{NW}, A_i^{SW}, A_i^{SE} and A_i^{NE} are set to zero in Eq. (B.55). This also leads to $a_i = u_i = A_i^{WW} = A_i^{EE} = 0$.

Therefore, for a 5-point formulation, Eq. (B.55) reduce to

$$b_i = B_i \tag{B.56a}$$

$$c_i f_{i-m+1\,i} = C_i \tag{B.56b}$$

$$b_i f_{i-m} + c_i = 0 \tag{B.56c}$$

$$b_i g_{i-m} + d_i = D_i \tag{B.56d}$$

$$b_i h_{i-m} + c_i g_{i-m+1} + d_i f_{i-1} + e_i = E_i \tag{B.56e}$$

$$c_i h_{i-m+1} + e_i f_i = F_i \tag{B.56f}$$

$$d_i g_{i-1} = G_i \tag{B.56g}$$

$$d_i h_{i-1} + e_i g_i = 0 \tag{B.56h}$$

$$e_i h_i = H_i \tag{B.56i}$$

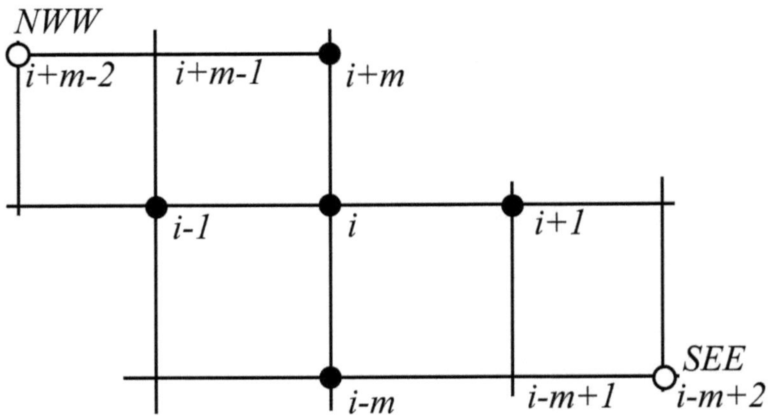

Fig. B.6 Numerical molecule for 5-point MSIP

The partial cancellation is done in a similar manner as described for SIP. Taylor series approximations for x_{i-m+2} and x_{i+m-2} are written as

$$x_{i-m+2} = -2x_i + 2x_{i+1} + x_{i-m} \tag{B.57}$$

and

$$x_{i+m-2} = -2x_i + 2x_{i-1} + x_{i+m} \tag{B.58}$$

After introducing a cancellation factor α in a similar fashion, the modified equations can be written as

$$\begin{aligned} B_i x_{i-m} + D_i x_{i-1} + E_i x_i + F_i x_{i+1} + H_i x_{i+1} \\ + C_i[x_{i-m+2} - \alpha(-2x_i + 2x_{i+1} + x_{i-m})] \\ + C_i[x_{i+m-2} - \alpha(-2x_i + 2x_{i-1} + x_{i+m})] = r_i \end{aligned} \tag{B.59}$$

Rearranging terms, Eq. (B.59) can be written as

$$\begin{aligned} (B_i - \alpha C_i)x_{i-m} + (D_i - 2\alpha G_i)x_{i-1} \\ + [E_i + 2\alpha(C_i + G_i)]x_i \\ + (F_i - 2\alpha C_i)x_{i+1} + (H_i - 2\alpha G_i)x_{i+m} \\ + C_i x_{i-m+2} \\ + C_i x_{i+m-2} = r_i \end{aligned} \tag{B.60}$$

Equating the corresponding terms in Eqs. (B.60) and (B.56) one obtains

$$b_i = B_i - \alpha C_i \tag{B.61a}$$

$$c_i f_{i-m+1} = C_i \tag{B.61b}$$

$$b_i f_{i-m} + c_i = 0 \tag{B.61c}$$

$$b_i g_{i-m} + d_i = D_i - 2\alpha G_i \tag{B.61d}$$

$$b_i h_{i-m} + c_i g_{i-m+1} + d_i f_{i-1} + e_i = E_i - 2\alpha(C_i + G_i) \tag{B.61e}$$

$$c_i h_{i-m+1} + e_i f_i = F_i - 2\alpha C_i \tag{B.61f}$$

$$d_i g_{i-1} = G_i \tag{B.61g}$$

$$d_i h_{i-1} + e_i g_i = 0 \tag{B.61h}$$

$$e_i h_i = H_i - \alpha G_i \tag{B.61i}$$

Algebraic manipulation of Eq. (B.61) gives

$$\begin{aligned} b_i &= B_i - \alpha c_i f_{i-m+1} \\ &= B_i - \alpha b_i f_{i-1} f_{i-m+1} \end{aligned} \tag{B.62a}$$

$$b_i f_{i-m} + c_i = 0 \tag{B.62b}$$

$$b_i g_{i-m} + d_i = D_i - 2\alpha d_i g_{i-1} \tag{B.62c}$$

$$b_i h_{i-1} + c_i g_{i-m+1} + d_i f_{i-1} + e_i = E_i + 2\alpha (c_i f_{i-m+1} + d_i g_{i-1}) \tag{B.62d}$$

$$c_i h_{i-m+1} + e_i f_i = F_i - 2\alpha c_i f_{i-m+1} \tag{B.62e}$$

$$d_i h_{i-1} + e_i g_i = 0 \tag{B.62f}$$

$$e_i h_i = H_i - \alpha d_{i-1} \tag{B.62g}$$

Collecting and rearranging terms once more gives the formulas for the coefficients of the strength vectors as follows:

$$b_i = B_i / (1 - \alpha f_{i-m} f_{i-m+1}) \tag{B.63a}$$

$$c_i = -b_i f_{i-m} \tag{B.63b}$$

$$d_i = (D_i - b_i g_{i-m}) / (1 + 2\alpha g_{i-1}) \tag{B.63c}$$

$$e_i = E_i - b_i h_{i-m} + c_i (2\alpha f_{i-m+1} - g_{i-m+1}) + d_i (2\alpha g_{i-1} - f_{i-1}) \tag{B.63d}$$

$$f_i = [F_i - c_i (h_{i-m+1} + 2\alpha f_{i-m+1})] / e_i \tag{B.63e}$$

$$g_i = -d_i h_{i-1} / e_i \tag{B.63f}$$

$$h_i = (H_i - \alpha d_i g_{i-1}) / e_i \tag{B.63g}$$

Once the factorization is completed, the iteration process described for the SIP method is applied in a similar fashion.

Appendix C
FORTRAN Subroutines—BLOCKSOLFICS2 and BLOCKSOLSICS

The following FORTRAN subroutines are a complement from the Author of this book to the readers.

Upon the input of the vectors from the block coupled system, the FICS-2 or the SICS process is started by calculating the necessary strength and defect vectors and continued with repeated inner iterations until convergence to a specified level of convergence specified.

The programs are not meant to be professional ones. They are directly extracted from the testing codes of the Author. The programs can be written in a more professional form in order to save computer's memory.

The subroutine RESIDUAL is common to both subroutines.

```
INPUT: AUS,AUW,AUP,AUE,AUN,AUS,PW,PE,BU: Coefficients of  u-momentum equations:
        AVS,AVW,AVP,AVE,AVN AVS,PS,PN,BV : Coefficients of  v-momentum equations:
                            UW,UE,VN,VS,BP: Coefficients of  mass conservation equations
                                         M: Number of grid points on one row
                                        NT: Total number of grid points
                              INMIN,INMAX: minimum and maximum number of iterations
                                      PREC: Precision to which iterations will be continued
                                     ALPHA: Relaxation parameter
OUTPUT:                              U,V,P: Solution (initially set in the main program)
                                      ITNO: Number of block iterations performed
```

© Springer International Publishing Switzerland 2016

Z. Mazhar, *Fully Implicit, Coupled Procedures in Computational Fluid Dynamics*,
Fluid Mechanics and Its Applications 115, DOI 10.1007/978-3-319-29895-5

SUBROUTINE FOR FICS-2

```
      SUBROUTINE BLOKSOLFICS2(AUS,AUW,AUP,AUE,AUN,AVS,AVW,AVP,AVE,AVN,PW
     1,PE,PS,PN,UE,UW,VN,VS,BU,BV,BP,NT,M,INMIN,INMAX
     2,PREC,ALPHA,U,V,P,ITNO)
      REAL*8   AUS(NT),AUW(NT),AUP(NT),AUE(NT),AUN(NT)
      REAL*8   AVS(NT),AVW(NT),AVP(NT),AVE(NT),AVN(NT)
      REAL*8   PW(NT),PE(NT),PS(NT),PN(NT)
      REAL*8   UW(NT),UE(NT),VS(NT),VN(NT)
      REAL*8   BU(NT),BV(NT),BP(NT)
      REAL*8   U(NT),V(NT),P(NT)
      REAL*8   A(NT),X(NT),B(NT),G(NT),S(NT),W(NT),C(NT),GA(NT),SI(NT)
      REAL*8   H(NT),F(NT),XK(NT),XN(NT)
      REAL*8   RHU(NT),RHV(NT),RHP(NT),YU(NT),YV(NT),YP(NT)
      REAL*8   UP(NT),VP(NT),PP(NT)
      REAL*8   DU(NT),DV(NT),DP(NT)
C
       CALL RESIDUAL(M,NT,AUS,AUW,AUP,AUE,AUN,AVS,AVW,AVP,
      1AVE,AVN,BU,BV,BP,U,V,P,UE,UW,VN,VS,PN,PS,PE,PW
      2,RU,RV,RP,RHU,RHV,RHP)
       UTAR=PREC
       VTAR=PREC
       PTAR=PREC
C       INCOMPLETE DECOMPOSITION
C      CALCULATE COEFFICIENTS (ONCE)
       DO I=1,NT
            H(I)=PW(I)/AUP(I)
            F(I)=PE(I)/AUP(I)
           XK(I)=PS(I)/AVP(I)
           XN(I)=PN(I)/AVP(I)
      ENDDO
C
         I=1
           X(I)=0.0
           B(I)=0.0
           G(I)=-(UE(I)*H(I)+VN(I)*XK(I))
           S(I)=-UE(I)*F(I)/G(I)
           W(I)=0.0
           C(I)=-VN(I)*XN(I)/G(I)
           GA(I)=0.0
           SI(I)=0.0
       DO  I=2,M-1
             X(I)=0.0
             B(I)=-UW(I)*H(I-1)
             G(I)=-(UW(I)*F(I-1)+UE(I)*H(I)+VN(I)*XK(I)+B(I)*S(I-1))
             S(I)=-UE(I)*F(I)/G(I)
             W(I)=-B(I)*C(I-1)/G(I)
             C(I)=-VN(I)*XN(I)/G(I)
             GA(I)=0.0
             SI(I)=-B(I)*W(I-1)
         ENDDO
```

```
      I=M
         X(I)=0.0
         B(I)=-UW(I)*H(I-1)
         G(I)=-(UW(I)*F(I-1)+UE(I)*H(I)+VN(I)*XK(I)+B(I)*S(I-1))
         S(I)=-UE(I)*F(I)/G(I)
         W(I)=-B(I)*C(I-1)/G(I)
         C(I)=-VN(I)*XN(I)/G(I)
         GA(I)=0.0
         SI(I)=-B(I)*W(I-1)
      DO I=M+1,NT
           A(I)=-VS(I)*XK(I-M)
           X(I)=-A(I)*S(I-M)
           GA(I)=-X(I)*S(I-M+1)
           B(I)=-(UW(I)*H(I-1)+A(I)*W(I-M))
           G(I)=-(UW(I)*F(I-1)+UE(I)*H(I)+VN(I)*XK(I)+VS(I)*XN(I-M))
           G(I)=G(I)-(A(I)*C(I-M)+X(I)*W(I-M-1)+B(I)*S(I-1))
           S(I)=-(UE(I)*F(I)+X(I)*C(I-M+1))/G(I)
           W(I)=-B(I)*C(I-1)/G(I)
           C(I)=-VN(I)*XN(I)/G(I)
           SI(I)=-B(I)*W(I-1)
      ENDDO
C
      ITNO=0
C
C     BEGIN INNER ITERATIONS
  200 CONTINUE
C
C      USE RHS=B-AX
C
C     CALCULATE YU, YV, YP (FORWARD SWEEP)
      DO I=1,NT
         YU(I)=RHU(I)/AUP(I)
         YV(I)=RHV(I)/AVP(I);ENDDO
         YP(1)=(RHP(1)-UE(1)*YU(1)-VN(1)*YV(1))/G(1)
      ENDDO

      DO  I=2,M
           YP(I)=(RHP(I)-UW(I)*YU(I-1)-UE(I)*YU(I)-VN(I)*YV(I)
     1     -B(I)*YP(I-1))/G(I)
      ENDDO

      DO I=M+1,NT
           YP(I)=(RHP(I)-UW(I)*YU(I-1)-UE(I)*YU(I)-VN(I)*YV(I)
     1          -VS(I)*YV(I-M)-A(I)*YP(I-M)-B(I)*YP(I-1)
     2          -X(I)*YP(I-M+1))/G(I)
      ENDDO

C
C     RESET REFERENCE PRESSURE
      DP(NT)=0.0
C     SOLVE FOR PRESSURE DIFFERENCE (BACKWARD SWEEP)
      DO I=NT-1,NT-M+2,-1
           DP(I)=YP(I)-S(I)*DP(I+1)
      ENDDO
```

```
      I=NT-M+1
      DP(I)=YP(I)-S(I)*DP(I+1)-W(I)*DP(I+M-1)
      DO I=NT-M,1,-1
          DP(I)=YP(I)-S(I)*DP(I+1)-C(I)*DP(I+M)-W(I)*DP(I+M-1)
      ENDDO

C     SOLVE FOR V-DIFFERENCE
      DO I=1,NT-M
          DV(I)=YV(I)-XK(I)*DP(I)-XN(I)*DP(I+M)
      ENDDO

      DO I=NT-M+1,NT
         DV(I)=YV(I)-XK(I)*DP(I)
      ENDDO

C     SOLVE FOR U-DIFFERENCE
      DO I=1,NT-1
           DU(I)=YU(I)-H(I)*DP(I)-F(I)*DP(I+1)
      ENDDO

      DU(NT)=YU(NT)-H(NT)*DP(NT)
C     UPDATE VELOCITY AND PRESSURES
      DO I=1,NT
          U(I)=U(I)+ALPHA*DU(I)
          V(I)=V(I)+ALPHA*DV(I)
          P(I)=P(I)+ALPHA*DP(I)
      ENDDO
C
      ITNO=ITNO+1
C
      CALL RESIDUAL(M,NT,AUS,AUW,AUP,AUE,AUN,
     1AVS,AVW,AVP,AVE,AVN,BU,BV,BP,U,V,P,UE,UW,VN,VS,
     2PN,PS,PE,PW,RU,RV,RP,RHU,RHV,RHP)
C
C     IF SATISFIED RETURN, ELSE CONTINUE ITERATING
C      RETURN IF RMS ERROR BELOW LIMITS
      IF(RU.LT.UTAR.AND.RV.LT.VTAR.AND.RP.LT.PTAR.AND.ITNO.GT.INMIN)
     1RETURN
C      RETURN IF MAX PERMITTED NUMBER OF ITERATIONS ARE PERFORMED
      IF(ITNO.GE.INMAX)RETURN
      GO TO 200
      END
```

SUBROUTINE FOR SICS

```
      SUBROUTINE BLOKSOLSICS(AUS,AUW,AUP,AUE,AUN,AVS,AVW,AVP,AVE,AVN,PW
     1,PE,PS,PN,UE,UW,VN,VS,BU,BV,BP,NTOT,IN,U,V,P,IIT,RMX,INMAX,XIP
     2,INO,PREC,ALPHA,INMIN)
      REAL*8   AUS(1600),AUW(1600),AUP(1600),AUE(1600)
      REAL*8   AUN(1600),AVS(1600),AVW(1600)
      REAL*8   AVP(1600),AVE(1600),AVN(1600),PW(1600)
      REAL*8   PE(1600),PS(1600),PN(1600)
      REAL*8   UE(1600),UW(1600),VN(1600),VS(1600),BU(1600)
      REAL*8   BV(1600),BP(1600),U(1600)
      REAL*8   V(1600),P(1600),S(1600)
      REAL*8   YU(1600),YV(1600),YP(1600)
      REAL*8   H(1600),F(1600),G(1600),XK(1600),RHU(1600),RHV(1600)
      REAL*8   RHP(1600),XN(1600)
      REAL*8   UP(1600),VP(1600),PP(1600)
      REAL*8   REU(1600),REV(1600),REP(1600)
      REAL*8   DU(1600),DV(1600),DP(1600),AL(1600)
      REAL*8   A(1600),B(1600),C(1600),D(1600),Y(1600)
      REAL*8   X(1600),W(1600),GA(1600),SI(1600)
C
      N=NTOT
C
C     FIRST CALCULATE INITIAL RMS LEVELS
C
      CALL RESIDUAL(IN,NTOT,AUS,AUW,AUP,AUE,AUN,AVS,AVW,AVP,
     1AVE,AVN,BU,BV,BP,U,V,P,UE,UW,VN,VS,PN,PS,PE,PW
     2,RU,RV,RP,REU,REV,REP)

       UTAR=PREC
       VTAR=PREC
       PTAR=PREC
C
C     INCOMPLETE DECOMPOSITION
C
      DO I=1,NTOT
         H(I)=PW(I)/AUP(I)
         F(I)=PE(I)/AUP(I)
         XK(I)=PS(I)/AVP(I)
         XN(I)=PN(I)/AVP(I)
      ENDDO

C
      I=1
        G(I)=-(UE(I)*H(I)+VN(I)*XK(I))

      DO I=2,IN
         G(I)=-(UW(I)*F(I-1)+UE(I)*H(I)+VN(I)*XK(I))
      ENDDO
```

```
      DO I=IN+1,NTOT
           G(I)=-(UW(I)*F(I-1)+UE(I)*H(I)+VN(I)*XK(I)+VS(I)*XN(I-IN))
      ENDDO

      IIP=0
      XIP=0.0

C     START INNER ITERATIONS

200 CONTINUE

      DO I=1,NTOT
           UP(I)=U(I)
           VP(I)=V(I)
           PP(I)=P(I)
      ENDDO
C
C      USE RHS=B-AX
C
      DO  I=1,NTOT
           RHU(I)=REU(I)
           RHV(I)=REV(I)
           RHP(I)=REP(I)
      ENDDO
C
C CALCULATE YU, YV, YP- FORWARD SWEEP
C
      DO I=1,NTOT
           YU(I)=RHU(I)/AUP(I)
           YV(I)=RHV(I)/AVP(I)
      ENDDO
C
      YP(1)=(RHP(1)-UE(1)*YU(1)-VN(1)*YV(1))/G(1)

      DO I=2,IN
          YP(I)=(RHP(I)-UW(I)*YU(I-1)-UE(I)*YU(I)-VN(I)*YV(I))/G(I)
      ENDDO

      NN=IN+1

      DO I=NN,NTOT
           YP(I)=(RHP(I)-UW(I)*YU(I-1)-UE(I)*YU(I)-VN(I)*YV(I)
     1      -VS(I)*YV(I-IN))/G(I)
      ENDDO

      PR=YP(NTOT)
      DP(NTOT)=0.0

C     CALCULATE DP, DV, DU- BACKWARD SWEEP

      DO I=NTOT-1,1,-1
           DP(I)=YP(I)-PR
      ENDDO
C
```

```
C
      NN=NTOT-IN
      DO  I=1,NN
          DV(I)=YV(I)-XK(I)*DP(I)-XN(I)*DP(I+IN)
      ENDDO

      NN=NTOT-IN+1
      DO I=NN,NTOT
         DV(I)=YV(I)-XK(I)*DP(I)
      ENDDO

      NN=NTOT-1
      DO I=1,NN
         DU(I)=YU(I)-H(I)*DP(I)-F(I)*DP(I+1)
      ENDDO

      DU(NTOT)=YU(NTOT)-H(NTOT)*DP(NTOT)

C     UPDATE U, V, P

      DO I=1,NTOT
         U(I)=UP(I)+ALPHA*DU(I)
         V(I)=VP(I)+ALPHA*DV(I)
         P(I)=PP(I)+ALPHA*DP(I)
      ENDDO
C
      IIP=IIP+1
      XIP=XIP+1.0
C
      CALL RESIDUAL(IN,NTOT,AUS,AUW,AUP,AUE,AUN,
     1AVS,AVW,AVP,AVE,AVN,BU,BV,BP,U,V,P,UE,UW,VN,VS,
     2PN,PS,PE,PW,RU,RV,RP,REU,REV,REP)
C
C      RETURN IF RMS ERROR BELOW LIMITS
C
C      WRITE(6,53)IIP,RU,RV,RP
   53 FORMAT(2X,'ITE=',I4,' RU=',F10.7,2X,'RV=',F10.7,2X,'RP=',F10.7)
      IIK=IIP+1
      IF(RU.LT.UTAR.AND.RV.LT.VTAR.AND.IIK.GT.INMIN)RETURN
C
C      RETURN IF MAX PERMITTED NUMBER OF ITERATIONS ARE PERFORMED
C
      IF(IIP.GE.IIT)RETURN
      GO TO 200
      END
```

SUBROUTINE RESIDUAL

```
      SUBROUTINE RESIDUAL(M,NT,AUS,AUW,AUP,AUE,AUN,
     1AVS,AVW,AVP,AVE,AVN,BU,BV,BP,U,V,P,UE,UW,VN,VS,
     2PN,PS,PE,PW,RU,RV,RP,REU,REV,REP)
      REAL*8   AUS(NT),AUW(NT),AUP(NT),AUE(NT);AUN(NT)
      REAL*8   AVS(NT),AVW(NT),AVP(NT),AVE(NT),AVN(NT)
      REAL*8   BU(NT),BV(NT),BP(NT),U(NT),V(NT),P(NT),UE(NT)
      REAL*8   UW(NT),VN(NT),VS(NT),PN(NT),PS(NT),PE(NT),PW(NT)
C
C     CALCULATE RESIDUALS
      DO I=1,NT
            REU(I)=BU(I)-AUP(I)*U(I)-PW(I)*P(I)
            REV(I)=BV(I)-AVP(I)*V(I)-PS(I)*P(I)
            REP(I)=BP(I)-UE(I)*U(I)-VN(I)*V(I)
      ENDDO

      DO I=2,NT
           REU(I)=REU(I)-AUW(I)*U(I-1)
           REV(I)=REV(I)-AVW(I)*V(I-1)
           REP(I)=REP(I)-UW(I)*U(I-1)
      ENDDO

      DO I=M+1,NT
          REU(I)=REU(I)-AUS(I)*U(I-M)
          REV(I)=REV(I)-AVS(I)*V(I-M)
          REP(I)=REP(I)-VS(I)*V(I-M)
      ENDDO

      DO I=1,NT-1
         REU(I)=REU(I)-AUE(I)*U(I+1)
         REV(I)=REV(I)-AVE(I)*V(I+1)
         REU(I)=REU(I)-PE(I)*P(I+1)
      ENDDO

      DO I=1,NT-M
         REU(I)=REU(I)-AUN(I)*U(I+M)
         REV(I)=REV(I)-AVN(I)*V(I+M)
         REV(I)=REV(I)-PN(I)*P(I+M)
      ENDDO

C     CALCULATE NORMALIZED RMS VALUES
      RU=0.0
      RV=0.0
      RP=0.0
      DO  I=1,NT
        BS=ABS(REU(I))
        RU=RU+BS*BS
        BS=ABS(REV(I))
        RV=RV+BS*BS
        BS=ABS(REP(I))
        RP=RP+BS*BS
     ENDDO
        RU=SQRT(RU)/NT
        RV=SQRT(RV)/NT
        RP=SQRT(RP)/NT
        RETURN
        END
```

References

1. Amsden AA, Harlow FH (1970) The SMAC method: A numerical technique for calculating incompressible fluid flows. Los Alamos Scientific Laboratory Report No. LA-4370, Los Alamos, New Mexico
2. Ashrafizadeh A, Alinia B, Mayeli P (2015) A new co-located pressure-based discretization method for the numerical solution of incompressible Navier-Stokes equations. Numer Heat Transf Part B: Fundam 67(6):563-589
3. Briley WR (1974) Numerical method for predicting three-dimensional steady viscous flow in ducts. J Comput Phys 14:8–28
4. Chalhub DJNM, Sphaier LA, de B. Alves LS (2014) Eliminating the pressure-velocity coupling from the incompressible Navier-Stokes Equations Using Integral Transforms. In: ICCM, Cambridge, England, 28-30 July 2014
5. Chatwani AU, Turan A (1991) improved pressure-velocity coupling algorithm based on minimization of global residual norm. Numer Heat Trans Part B 20:115–123
6. Choi HG, Yoo JY (1995) A hybrid numerical method for Navier-Stokes equations based on SIMPLE algorithm. Numer Heat Transf Part B 28:155–170
7. Chorin AJ (1967) A numerical method for solving incompressible viscous flow problems. J Comput Phys 2:12–26
8. Chorin AJ (1971) Numerical solution of the Navier-Stokes equations. Math Comput 22:745-762
9 Chow WK and Cheung YL (1997) Comparison of the Algorithms PISO and SIMPLER for Solving Pressure Linked Equations in Simulating Compartmental Fire. Numer Heat Trans Part A, vol 31, pp 87–112
10. Connell SD, Stow P (1986) A discussion and comparison of numerical techniques used to solve the Navier-Stokes and Euler equations. Int J Numer Methods Fluids 6:155–163
11. Darwish M, Moukallad F (2014) A fully coupled Navier-Stokes solver for fluid flow at all speeds. Numer Heat Transf Part B 65(5):410-444
12. Darwish M, Abdel Aziz A, Moukallad F (2015) A coupled pressure-based finite-volume solver for incompressible two-phase flow. Numer Heat Transf Part B 67(1):47–74
13. Deng GB, Piquet J, Queutey P, Visonneau M (1994) A new fully coupled solution of the Navier-Stokes equations. Int J Numer Methods Fluids 19:605–639
14. Freund RW, Nachtigal NM (1991) QMR: A quasi-minimal residual method for non-hermitian linear systems. Numer Math 60:315–339
15. Galpin PF, Van Doormaal JP, Raithby GD (1985) Solution of the incompressible mass and momentum equations by application of a coupled equation line solver. Int J Numer Methods Fluids 5:615–625
16. Galpin PF, Raithby GD (1986) A treatment of non-linearities in the numerical solution of the incompressible Navier-Stokes equations. Int J Numer Methods Fluids 6:409–426
17. Güngörmüş T (1995) A block implicit method for incompressible fluid flow problems. Master's Thesis, Eastern Mediterranean University, Turkish Republic of North Cyprus

© Springer International Publishing Switzerland 2016

Z. Mazhar, *Fully Implicit, Coupled Procedures in Computational Fluid Dynamics*,
Fluid Mechanics and Its Applications 115, DOI 10.1007/978-3-319-29895-5

18. Hanby RF, Silvester DJ, Chow JW (1996) A comparison of coupled and segregated iterative solution techniques for incompressible swirling flows. Int J Numer Methods Fluids 22:353–373
19. Harlow FH, Welch JE (1965) Numerical calculation of time-dependent viscous incompressible flow of fluid with free surface. Phys Fluids 8(12):2182–2189
20. Henau VD, Raithby GD, Thompson BE (1989) A total pressure correction for upstream weighted schemes. Int J Numer Methods Fluids 9:855–864
21. Issa RI (1986) Solution of implicitly discretized fluid flow equations by operator-splitting. J Comput Phys 62:40–65
22. Jennings A (1977) Matrix computations for engineers and scientists, John Wiley and Sons Ltd
23. Kim CJ, Ro ST (1995) a block correction-aided, strongly implicit procedure to treat simultaneous linear equations arising from implicit discretization of three-dimensional field problems. Numer Heat Transf Part B 28:371–384
24. Lage PLC (1996) Modified strongly implicit procedure with adaptive optimization of its iteration parameter. Numer Heat Transf Part B 30:255–270
25. Latimer BR, Pollard A (1985) Comparison of pressure-velocity coupling solution algorithms. Numer Heat Transf 4:409–425
26. Mangani L, Buchmayr M, Darwish M (2014) Development of a fully coupled solver in OpenFOAM: Steady-state incompressible turbulent flows. Numer Heat Transf Part B 6(1):1–20
27. Mazhar Z, Raithby GD (1981) A refined PUMPIN (Pressure Update by Multiple Path Integration) method for updating pressures in the numerical solution of the incompressible fluid flow equations. In: Proceedings 2nd international conference on numerical methods in laminar and turbulent flow. Pineridge Press, Swansea, UK, pp 255–266
28. Mazhar Z (1981) An evaluation of the segregated solution procedures for the solution of incompressible fluid flow problems. Ph.D. Thesis, University of Waterloo, Canada
29. Mazhar Z (2001) A procedure for the treatment of the velocity-pressure coupling problem in incompressible fluid flow. Numer Heat Transf Part B 39:91–100
30. Mazhar Z (2002) An enhancement to the block implicit procedure for the treatment of the velocity-pressure coupling problem in incompressible fluid flow. Numer Heat Transf Part B 41:493–500
31. Mazhar Z (2016) A novel fully implicit block coupled procedure for the ultimate treatment of the velocity-pressure coupling problem in incompresible fluid flow. Numer Heat Transf Part B accepted for publication
32. Patankar SV (1981) A calculation procedure for two-dimensional elliptic situations. Numer Heat Transf 4:409–425
33. Patankar SV (1974) A calculation procedure for the transient and steady-state behavior of shell-and-tube heat exchangers, chap. 7. In: Heat exchangers: Design and theory sourcebook. Hemisphere, Washington, DC
34. Patankar SV, Spalding DB (1970) Heat and mass transfer in boundary layers. Intertext, London
35. Patankar SV, Spalding DB (1972) A calculation procedure for heat, mass and momentum transfer in three-dimensional parabolic flows. Int J Heat Mass Transf 2:1782–1806
36. Patankar SV (1980) Numerical heat transfer and fluid flow. Hemisphere Publishing
37. Raithby GD, Torrance KE (1974) Upstream-Weighted differencing schemes and their application to elliptic problems involving fluid flow. Comput Fluids 2:191–206
38. Raithby GD, Schneider GE (1979) Numerical solution of problems in incompressible fluid flow: Treatment of the velocity-pressure coupling. Numer Heat Transf 2:417–440
39. Raithby GD (1976) A critical evaluation of upstream differencing applied to problems involving fluid flow. Comput Methods Appl Mech Eng 9:75–103

40. Runchal AK, Spalding DB, Wolfhstein M (1968) numerical solution of the elliptic equations for transport of vorticity, heat and matter in two-dimensional flow. Phys Fluids 12(Suppl II): (II-21)–(II-28)
41. Sathyamurty PS, Patankar SV (1994) Block-correction-based multigrid method for fluid flow problems. Numer Heat Transf Part B 25:375–394
42. Schneider GE, Zedan MG (1981) A modified strongly implicit procedure for the numerical solution of field problems. J Numer Heat Transf 4:1–19
43. Sheen SC, Wu JL (1997) Solution of the pressure correction equation by the preconditioned conjugate gradient method. Numer Heat Transf Part B 32:215–230
44. Stone HL (1968) Iterative solution of implicit approximations of multi-dimensional partial differential equations. SIAM J Numer Anal 3:530–558
45. Theodossiou VM, Sousa ACM (1986) An efficient algorithm for solving the incompressible fluid flow equations. Int J Numer Methods Fluids 6:557–572
46. Torrance KE, Rockett JA (1969) Convection in an enclosure with localized heating from below. J Fluid Mech 36:33–54
47. Van Doormaal JP, Raithby GD (1984) Enhancements of the SIMPLE method for predicting incompressible fluid flow. Numer Heat Transf 7:147–163
48. Varga RS (1962) Matrix Iterative analysis, Prentice Hall, Inc, Englewood Cliffs, NJ.
49. Wanik A, Schnell U (1989) Some remarks of the PISO and SIMPLE algorithms for steady turbulent flow problems. Comput Fluids 4(4):555–570
50. Wen X, Ingham DB (1994) A numerical method for accelerating the rate of convergence of the SIMPLE-like algorithm for flow through a thin filter. Int J Numer Methods Fluids 19:889–903
51. Yin R, Chow WK (2003) Comparison of four algorithms for solving pressure-velocity linked equations in simulating atrium fire. Int J Architect Sci 4(1):24–35
52. Zedan M, Schneider GE (1985) A coupled strongly implicit procedure for velocity and pressure computations in fluid flow problems. Numer Heat Transf 8:537–557
53. Zedan M, Schneider GE (1983) A strongly implicit simultaneous variable solution procedure for velocity and pressure in fluid flow problems. In: AIAA-83-1569, AIAA 18th thermophysics conference

Index

A
Algorithm, 83
Analytical solution, 28
Approximation, 9

B
Backward facing step, 99
Backward substitution, 17
Backward sweep, 17
BASIP, 136
Benchmark problem, 89
BIPEN, 80
Block-correction algorithm, 136
Block implicit decomposition, 23
Block incomplete decomposition, 65
Block matrix, 57
Block relaxation parameter, 100
Block solution vector, 62
Boundary condition, 41
  types of, 43

C
CELS, 135
Central difference, 10
Complexity, 82
Continuity equation, 19
Continuous, 9
Control volume, 27
Convergence, 22
CSIP, 135

D
Defect matrix, 69
Defect vector, 69
Dense shelter, 94
Density, 20, 43
Diagonal dominancy, 63
Diagonal vector, 17
Dirichlet, 44
Discretization, 4, 7, 8, 23, 25, 27, 31, 32, 34, 37, 73, 116, 119, 121, 129, 131, 133, 134, 141
Divergence, 9, 132
Downstream, 43
Durability, 102

E
E-factor formulation, 55
Eddy diffusivity, 20
Elliptic, 22
Equidistant spacing, 10
Exact solution, 90
Exponential Scheme, 130

F
Far-diagonal, 18
FICS-1, 80
FICS-2, 80
FIMOSE, 134
Finite difference approximation, 7
Flow geometry, 99
Flow pattern, 90
Flow separation, 94
Fluid flow, 19
Forward elimination, 15
Forward sweep, 17
Free surface, 43
Friction velocity, 43
Fully implicit coupled solution procedure, 5
Fundamental theorem of linear algebra, 7

G
Gaussian elimination, 62
Governing differential equations, 7, 19
Gradient type boundary condition, 46
Grid arrangement, 8
Grid aspect ratio, 102
Grid refinement, 95

© Springer International Publishing Switzerland 2016

Z. Mazhar, *Fully Implicit, Coupled Procedures in Computational Fluid Dynamics*,
Fluid Mechanics and Its Applications 115, DOI 10.1007/978-3-319-29895-5

Grid structure, 26
Grid system, 26

H
Heat transfer, 119
Hybrid scheme, 130

I
Incompressible fluid, 19
Inflow, 43
Inlet, 51
Inner iterations, 83, 100
Interpolation, 39
Iteration, 32

L
Lagrangian polynomial, 8
Linearisation, 20
Loosely diagonally dominant, 143
Lower triangular, 16

M
Mass conservation equation, 19
Mass control volumes, 42
Mass preservation, 20
Maximum pressure differences test, 71
MCGS, 136
Mean motion, 19
Momentum equations, 19
MSIP, 134
Multiplicative operation, 100

N
Neumann, 22, 44
NEW-PUMPIN, 133
Non-equidistant spacing, 10
Nonlinearity, 20
Normalized head, 78
No-slip, 52
Numbering scheme, 57
Numbering strategy, 57
Numerical simulation, 9
Numerical solution procedure, 26

O
Obstacle, 43
Off-diagonal vectors, 16
Optimal relaxation parameters, 101
Outflow, 43
Overall iteration, 20, 78
Overflow, 30
Over-relaxation, 56

P
Parabola, 7
Peclet number, 90
Performance, 99
PISO, 135
Pivoting, 63
Poisson, 22
Power-Law scheme, 130
Pressure, 20
Pressure correction, 63
Pressure difference, 20
Primitive form, 59
Profile assumption, 27
Programming strategy, 83
Programming technique, 82
PULS, 132
PUMPIN, 133

Q
QMR, 136
Quadratic interpolation, 7

R
Range of convergent parameters, 101
Reattachment point, 94
Recirculating flow, 99
Reference point, 64
Relativeness, of pressure, 21
Relative performance, 104
Relaxation, 55
Relaxation factor, 55
Residual, 71
Roughness parameter, 43
Running indices, 81

S
Segregated solution procedure, 63
Shelterbelt, 94
SIMPLE, 132, 141
SIMPLEC, 134
SIMPLER, 133, 141
SIP, 134
Solution domain, 26
Space coordinate, 20
Staggered grid, 27, 130
Steady flow, 19
Stoke's flow, 119
Storage configuration, 27
Storage requirements, 82
Strength matrix, 69
Strength vector, 69
Strictly diagonally dominant, 144
Successive iteration, 70

T
Taylor series, 7
Testing criterion, 99
Three-dimensional problem, 120
Time marching, 116
Tunnel, 99
Turbulence, 119

U
Under-relaxation, 55
Upper triangular, 16, 18
Upstream, 94
Upstream differencing, 130

V
Velocity- pressure coupling problem, 21
Velocity correction step, 144
Von Karman's constant, 43
Vorticity, 132

W
Weighted approximation, 39
Weighted averaging, 39
Weighting factor, 30
Wind-break, 43
Windbreak 52

MIX
Papier aus verantwortungsvollen Quellen
Paper from responsible sources
FSC® C105338

If you have any concerns about our products,
you can contact us on
ProductSafety@springernature.com

In case Publisher is established outside the EU,
the EU authorized representative is:
Springer Nature Customer Service Center GmbH
Europaplatz 3, 69115 Heidelberg, Germany

Printed by Libri Plureos GmbH
in Hamburg, Germany